The Solar System

THE SOLAR SYSTEM

The Sun, Planets, and Life

Roman Smoluchowski

SCIENTIFIC AMERICAN LIBRARY

An imprint of Scientific American Books, Inc.
New York

Library of Congress Cataloging in Publication Data:

Smoluchowski, Roman
    The solar system.

    (Scientific American library)
    Includes index.
        1. Solar system—Popular works.   2. Life on
    other planets—Popular works.   I. Title.
        QB501.2.S58   1983   523.2   83–11661
ISBN 0-7167-1492-2
ISBN 0-7167-1493-0 (pbk.)

Printed in the United States of America

Scientific American Library is published by
Scientific American Books, Inc., a subsidiary
of Scientific American, Inc.

Distributed by W.H. Freeman and Company,
41 Madison Avenue, New York, New York 10010.

2 3 4 5 6 7 8 9 0 KP 1 0 8 9 8 7 6 5 4

With love to Louise

CONTENTS

We live in an old chaos of the sun,
Or old dependency of day and night,
Or island solitude, unsponsored, free,
Of that wide water, inescapable.
Sunday Morning, WALLACE STEVENS

PREFACE

It is difficult to imagine a reasonably intelligent human being who has not wondered about the cosmos, our place in it, and its origins. Most of us, however, are unable to devote much time to such thoughts, and it was to help remedy this situation that I wrote this book, which offers a quick glance at the Sun and the planets and their moons and assesses the chances that there is life elsewhere in the solar system. I have done this using a minimum of abstract terminology and no mathematics, leaning instead on visual material.

It is easy to be moved simply to contemplation by the beauty and grandeur of the world around us and by its still uncounted mysteries. Specific questions do come to mind, however—first, those concerned with how it all began and, particularly, how the Sun and its planetary family originated. Intuition tells us that there was probably a beginning and that there must be an end, but it is the knowledge accumulated over the centuries and now augmented by space exploration that provides the real answers. Though the spectacular successes of American and Soviet space exploration have greatly stimulated interest in the planets, it is important to realize that the fascinating images obtained by these spacecraft are only a fraction of the data collected during these missions. The images have an immediate visual appeal, but the significance of the other data becomes evident only if one has a certain amount of background information. This book aims to give its reader this kind of information—first about the formation of the Sun in the interstellar chaos, then about the Sun itself, about the planets and their formation, the innumerable small bodies, and finally, the ultimate death of the Sun. An afterword addresses the difficult problems of the origin of life and of whether we are or are not alone, whether there are beings other than ourselves pondering the mysteries of the desolate immensity of the universe.

I am greatly indebted to my former Princeton colleagues—Robert H. Dicke, John J. Hopfield, P. J. E. Peebles, and the late Robert E. Danielson—for helping me to enter the world of planetology, to my present colleagues in the Astronomy and Physics Departments at the University of Texas for letting me share their enthusiasm and to John A. Wheeler—at Princeton and now in Austin—for suggesting that I write this book for the SCIENTIFIC AMERICAN LIBRARY. In particular I want to thank my wife Louise for her unlimited support and for scrutinizing the text for clarity and style.

Roman Smoluchowski
Austin, Texas
May, 1983

The Solar System

Facing page: *The birth of the Sun, in an artist's rendering.*

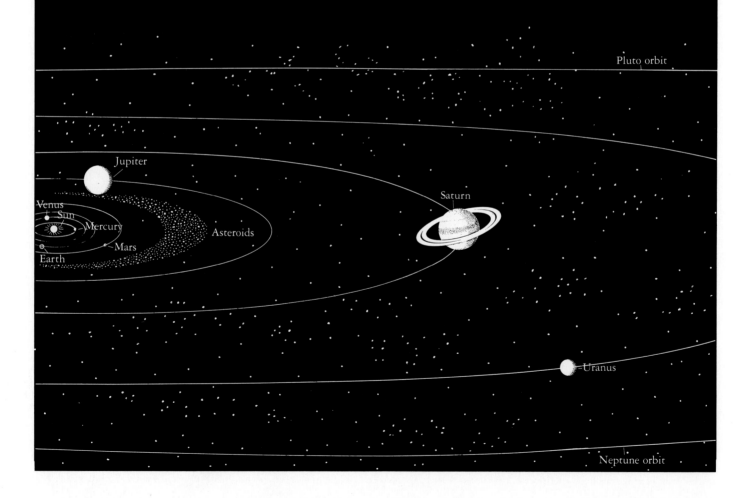

Even though the diagram of the solar system below is drawn at a very small scale, the distances involved are so great that it is not possible to show the whole system within the limits of the page. To convey some impression of this enormousness we have taken the text of this book to represent the distance from the Sun to the outermost planetary orbit, and have distributed small symbols of the various bodies (at the heads of pages) in their relative positions. Thus, we begin with the Sun here (above), Mercury appears on page 2, and so on, with Pluto ending the sequence on page 158.

Pluto orbit

Jupiter

Saturn

Venus
Sun
Mercury
Asteroids

Mars

Earth

Uranus

Neptune orbit

ESSAY I   # THE BIRTH OF THE SUN

Our Sun, with its planets, is about three-fifths of the way from the center of a spiral galaxy that we see in all directions, and, in particular, in cross-section as the Milky Way. With one exception, all the stars and other objects that can be seen with the naked eye are part of our galaxy. The exception is the spectacular Andromeda galaxy, which reveals its structure when observed even with a small telescope. This galaxy is huge—light takes more than 100,000 years to cross from one edge of its disk to the other—but to a naked eye it is just a barely visible small blur. The rest of the universe, the millions of other galaxies, only a fraction of which are spiral, can be seen only with the use of powerful telescopes. Our Galaxy, far from the largest, contains as many as 100 *billion* stars.

Because we are inside our galaxy, our line of sight along its plane is obstructed in most directions by gas and dust, making it difficult to investigate. For this reason the center of our galaxy is not observa-ble by means of visible light, and we are only just beginning to learn about this region by observing radiation of a kind that does penetrate the dust. Layers of dust similar to those in our galaxy can be seen in certain galaxies that we happen to see nearly edge on. To understand the evolution of our Sun and of its planetary system it is therefore helpful to study the conditions in galaxies that can be seen by looking away from the dust- and murk-hidden plane of our own galaxy.

Galaxies contain huge clouds of gas, as well as stars, and these move not only around the galactic center, but also slowly rotate or spin and wander through the galaxy itself. Thus, a galaxy is never quiet; there is always motion: our Sun, for example, together with its planets is slowly drifting in the direction of the star Vega. If a galaxy cloud—predominantly hydrogen and helium and very fine dust—attains a mass of more than a few thousand times that of our Sun and if it is not above a certain critical temperature, it may col-

The Andromeda galaxy, although over 2 million light years away, is the nearest one to our galaxy, the Milky Way. Note the huge, central, very hot condensation, the thinner disk, and two curved arms.

Above right, the Whirlpool galaxy, presumably similar to the Milky Way, displays the central condensation and arms typical of a spiral galaxy. Spiral galaxies are the largest known objects in the universe that have an obvious structure.

lapse—fall together—because of the mutual gravitational attraction of its parts. This collapse of a cloud is the initial stage of the long process of the formation of all stars, including our Sun. Although a small or hot gas cloud may disperse because its internal gravitational attraction cannot overcome the outward pressure of its hot gases, which try to escape like steam from a kettle, there are circumstances in which it can collapse. In one such case a relatively small galactic gas cloud—one of mass, say, a few hundred times that of the Sun—might enter one of the spiral arms of the galaxy, where, in contrast to the rather thinly populated space between the arms, there are many large stars and other matter, which produce high gravitational forces that may induce the small cloud to collapse. Another possibility is the rather frequently observed phenomenon of a supernova— the explosion of an old decaying star— near the cloud, the huge shock wave from which triggers the collapse of the cloud, somewhat as a sound wave can start an avalanche. The material thrown out into space in supernova explosions contains elements not present in the early galactic cloud, including radioactive ones.

Edge-on view of a spiral galaxy in Coma Berenice.

Supernova (arrow) that occurred in a nearby galaxy in 1959. Note huge increase in brightness over corresponding image in upper photograph, taken a few days earlier.

A rotating gaseous cloud (a) *gradually will develop, depending on conditions, a ring or spiral arms and a central object* (b).

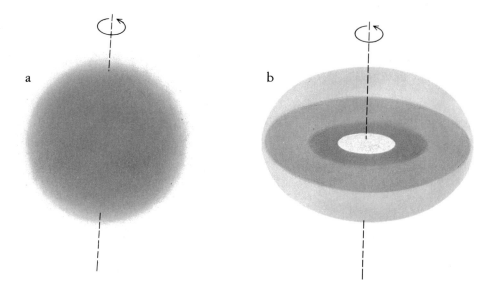

a   b

That our Sun and the solar system appear to have had a considerable concentration of these short-lived radioactive elements suggests that the collapse of the cloud from which our Sun and we were made was indeed initiated by a supernova.

How did our Sun form in such a collapsed cloud? The temperature of a shrinking cloud rises, especially in the central region, just as it does in any gas that is rapidly compressed—as for instance in a tire being inflated. While collapsing, a rotating cloud may break up into two or more parts each having the highest temperature close to its center. Recent studies of the process of the collapse of a cloud, by Peter Bodenheimer, of the Lick Observatory, reveal important details of the fragmentation process: If the rotating cloud was cold and had an initial spherical shape it would form a symmetrical disk. If, however, its temperature were higher, the collapse could be slightly irregular, not exactly circularly symmetrical, and the resulting disk would have a ring-like structure with two or more clumps of matter in it.

In a year, a cloud can shrink from an initial diameter of 2000 billion kilometers to a diameter of 200 million kilometers, its central temperature rising at the same time to 5,000 K. For a cloud comparable in mass to our Sun, 10 million years of such contraction will heat its center to a temperature of 10 million degrees K. (The more massive the star, the higher the temperature its center will reach.) At this important moment there begins a nuclear fusion reaction in which hydrogen is transformed into helium and that can be taken as the moment of the birth of a star. In this reaction, nuclei of four hydrogen atoms unite to form the nucleus of a helium atom, 99.3 percent of the mass of the four hydrogen nuclei going to make up the mass of the helium nucleus; the remaining 0.7 percent changes into energy in a conversion that obeys Einstein's famous $E = mc^2$, which is derived from relativity theory and in which m is the

Gaseous nebula in the constellation Orion. The small, dark regions are believed to be protostars—pockets of condensed gas from which new stars will ultimately form. The young stars illuminate and excite the colder gas.

mass loss and c is the speed of light. In such a reaction the amount of energy emitted by only 300 kilograms of hydrogen would equal the total energy consumed in the United States per day. This is about one tenth the amount of hydrogen needed to fill the famous dirigible Graf Zeppelin. To gain the same energy by simply burning hydrogen in air one would have to use 10,000 times this amount.

The hydrogen-to-helium and other similar nuclear-fusion reactions are the source of the energy—the heat and light—that makes the stars visible and our Sun shine so brightly. The basic hydrogen-to-helium reaction may raise the temperature of the star to the point where a new reaction, one that transforms helium to carbon, begins. If the temperature increases further, then another reaction, in which carbon is transformed to oxygen, may begin, and so on up the ladder of chemical complexity.

Once nuclear reactions have been initi-ated in its core, a star will exist in a steady state as a self-contained, extremely bright object, maintaining its temperature at a fairly constant level and gradually consuming its supply of hydrogen and of other elements. There are areas in the arms of galaxies where the formation of new stars is known to be particularly intense—such as in the Orion nebula.

The size and the energy production of our Sun indicate that it was formed in one of the arms of our Galaxy some 5 billion years ago. Since then it has made about 20 trips around the galactic center at a velocity of 250 kilometers per second, which is about one thousand times faster than the speed of sound on Earth. Presumably it will exist in more or less the same condition and place for another 5 billion years—almost a million times the period of recorded human history. The changes that will follow this period—and ultimately the death of the Sun—are discussed in the last essay, "The Fate of the Solar System," which begins on p. 139.

CHAPTER 1 THE SUN

CHAPTER 1   THE SUN

All life as we know it depends on the Sun. Its light and heat play an essential role in many ways. Photosynthesis, the life process in green plants, and ultimately the source of all food as well as coal and oil, is powered by sunlight. The seasons, the circulation of the air, the formation of clouds and rain—all occur as a direct result of the influence of the Sun. In view of its importance in our existence it is natural to try to learn whatever we can about its fundamental nature.

The Sun is located, as we have said, in one of the arms of our galaxy (The Milky Way), three-fifths of the way out from the galactic center. Formed there some 5 billion years ago from a cloud of gas, it has since made about 20 trips around the galactic center at a velocity of 250 kilometers per second. The sun is a very ordinary star; in our galaxy, which contains around 100 billion stars, there must be millions of others approximately the same size and temperature as ours—and there are millions and millions of galaxies in the universe.

The visible disk of the Sun has a diameter of about 1,400,000 kilometers, which is more than 100 times that of the Earth, and its gravitational pull is about 27 times stronger. All the various chemical elements that exist on the Earth exist also in the Sun, but in the latter everything is gaseous—that is, not liquid or solid. For this reason the Sun has no surface in the conventional sense, and what we refer to as its surface is simply the layer that emits light and that we can see; this layer is called the photosphere.

Preceding page: Radiotelescope image of the brightness distribution of the Sun. The red, yellow, and white areas reflect levels of intensity (with white most active) and correspond to regions in the corona that lie above sunspots.

Layer of dust in the plane of our galaxy, visible here in a view toward the center of the Milky Way, makes it difficult to explore directly our galaxy's distant parts.

Schematic representation of our Milky Way galaxy edge-on, showing the position of the Sun in it. The diameter of the galaxy is about 100,000 light years. It is surrounded by a huge but thin spherical halo of stars, some of which form clusters.

As in all stars, the conversion of hydrogen to helium, and other similar nuclear fusion reactions, known from experiments in laboratories, are the sources of the Sun's energy, heat, and light. The hydrogen-to-helium reaction is essentially the same one that occurs in an explosive manner in hydrogen bombs.

The Sun appears quite different according to when and how it is observed—through a telescope on a clear day, for example, or during a full solar eclipse, or photographed in the light emitted by hot hydrogen on its surface or by means of the X-rays it emits.

As with any star, there is clearly much more to the Sun than just a uniform ball of very hot gas. Analysis of the light coming from the Sun tells us the surface composition, the surface temperature (about 6000 K), and the rate of energy emission in all directions. The latter is enormous and is equivalent to the energy that would be generated by 100 billion tons of TNT exploding *per second.* Actually the Sun emits per second, in all directions, as much energy as Earth receives from it in 100 years. From these and other data, including solar mass (determined from the motion of the planets) and size, one can construct models of the solar interior. Such models indicate that, at the center, the temperature is about 16 million degrees K, the pressure about one billion times atmospheric pressure on Earth, and density about 160 times that of water. The central part is so dense that about half the mass of the Sun is contained in 1.6 percent of its volume! Energy production is similarly concentrated: As discussed earlier, the basic nuclear reaction in which hydrogen is transformed into helium, and which fuels the Sun, requires temperatures higher than ten million degrees; as a result,

The Sun in white light, with sunspots
visible (a), in eclipse (b), as seen in
hydrogen-alpha radiation (c), and in an
X-ray photograph (d).

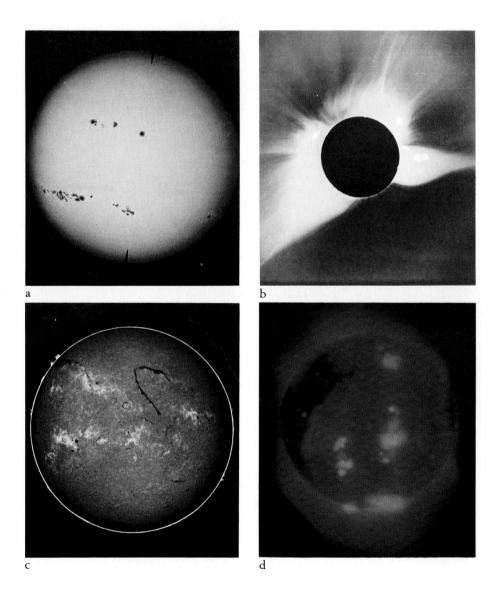

all solar nuclear energy is produced within the inner quarter of its radius, which
encloses about 1.6 percent of its volume. The enormous energy emitted by the
Sun requires the conversion (or "burning") of 650 million tons of hydrogen into
645.5 million tons of helium per second, or a net loss of mass of 4.55 million
tons per second. This does not, however, imply the imminent exhaustion of the

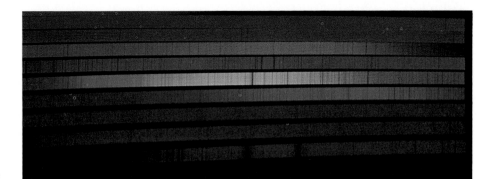

The solar spectrum, produced by passing the Sun's light through a prism. The various colors and lines indicate the presence of the elements of which it is composed. Lines here show that the Sun contains, among many other elements, hydrogen, iron, sodium, and calcium.

Apparatus for detection of solar neutrinos, located almost 1500 meters below the surface in Lead, South Dakota. The tank contains 400,000 liters of perchloroethylene, normally used in dry-cleaning.

Sun, for its mass is 100 billion billion times this loss!* In fact, even after about 5 billion years, when the Sun will begin to show the first signs of old age and start its transformation into a red giant, it will have converted only one thousandth of its original mass into energy.

Although we think we understand reasonably well the source of the Sun's energy, results of an important recent study seem to be difficult to reconcile with accepted views. The difficulty lies in the so-called neutrino problem. The nuclear reactions taking place in the Sun's center are accompanied by the emission of fast neutrinos, rather peculiar subatomic particles that have no mass and no charge and so can interact only very weakly with matter. As a consequence, they travel rapidly and virtually unimpeded from the center of the Sun to the Earth in about eight minutes. Delicate experiments performed in deep mines, to exclude other radiation, showed the flow of these solar neutrinos to be substantially smaller than what one would expect from the rate of heat emission of the Sun and thus the rate of the nuclear reactions in its center. Many proposals have been made to explain these results but the question is still unresolved, although it seems certain that the answer will be related to an energy flow slightly different from that now assumed and to a slightly lower temperature in the Sun's center.

What keeps the rate of energy production of the Sun so constant? Why doesn't it, like a giant nuclear reactor, explode, with catastrophic results for life on Earth? The answer lies in part in a kind of built-in self-regulator or thermostat: When, in the innermost part of the Sun, the hydrogen-to-helium reaction exceeds a certain rate, temperature increases there, causing the gases to expand. Because the gases expand, the temperature drops, producing a slowing down of the nuclear reaction. Nevertheless, there is evidence that, over billions of years, the energy emission of the Sun has dropped some 20 percent. In spite of this drop biologists believe that the approximately constant mean annual temperature on the Earth played an absolutely essential role in the evolution of intelligent life.

---

*Figures of this magnitude are often expressed in exponential form to the base 10, a convenient shorthand device. Thus the figure just cited—100 billion billion—would be expressed as 10^{20}. *Powers of Ten,* first book of the series in which this is the sixth, uses this exponential system as a framework for its exploration of the macro and micro worlds of nature.

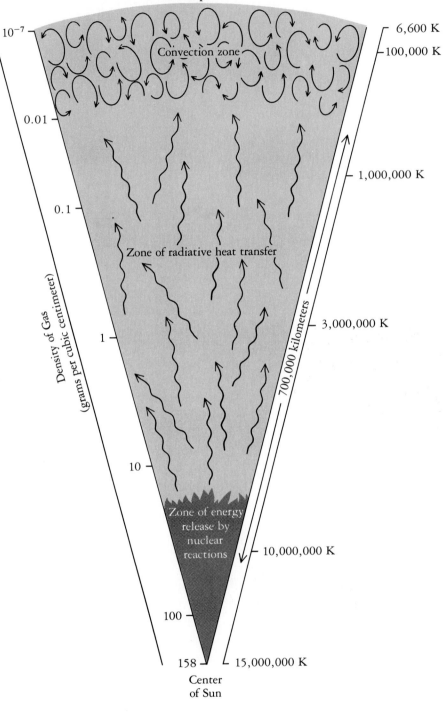

Photosphere

10⁻⁷ — 10^{-7}

Convection zone

6,600 K

100,000 K

0.01

0.1

1,000,000 K

Zone of radiative heat transfer

3,000,000 K

1

10

Zone of energy release by nuclear reactions

10,000,000 K

100

158

15,000,000 K

Center of Sun

Density of Gas (grams per cubic centimeter)

700,000 kilometers

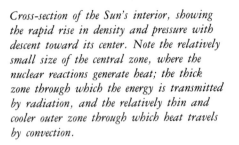

Cross-section of the Sun's interior, showing the rapid rise in density and pressure with descent toward its center. Note the relatively small size of the central zone, where the nuclear reactions generate heat; the thick zone through which the energy is transmitted by radiation, and the relatively thin and cooler outer zone through which heat travels by convection.

As described above, the region in which most of the solar energy is produced is virtually a pinpoint relative to the whole Sun. The manner in which the enormous energy reaches the solar surface—and eventually the Earth and other planets—also contributes to the constancy of the rate of radiation of energy. Between the very small central part of the Sun, where energy is produced, out to about three-quarters of its radius, the energy is transported by radiation. This means that light emitted by an atom in one place is absorbed by another atom some distance away and that, after a very short time, say one-billionth of a second, it is reradiated in some random direction, where it is reabsorbed, reradiated, and so on. This haphazard, undirected process of absorption and re-emission of light (that is, energy), gradually brings the energy to within one-quarter of the solar radius from the surface, where the temperature is about one-tenth that at the center. There the mode of energy transport changes from radiation to convection, which, being directional and relatively efficient, brings the energy to the solar surface within a day or two. (Convection is the familiar process of mixing that occurs in a gas or liquid heated in a non-uniform way—as in the case of a pot of heated soup. The existence of this convection layer is nicely confirmed by studies of solar pulsations, described later.) Nevertheless, in spite of the rapidity of the final step, it takes a particular unit of light or energy *one million years* to travel from the Sun's center to the surface. If, instead of being continuously absorbed and re-emitted in random directions, the light could go directly and freely from the Sun's central part of the surface, it would reach it in just over two seconds. The solar surface radiates energy freely into space, and that energy reaches us about eight minutes after it leaves the Sun. This remarkable slowness of the energy flow in the Sun has a great advantage for us and for all life in that it assures a steady energy output, minimizing any large variations that might occur at the center.

That convection is limited to the outer one-quarter of the solar radius has a noteworthy consequence: As the hydrogen in the small central volume of the Sun is transformed into helium, essentially no fresh hydrogen drifts in towards the center from the huge nearby supply. The effect is not unlike a prairie fire, with the conflagration advancing slowly but steadily as it consumes its fuel. When the hydrogen near the solar center is depleted the Sun will become an old star and will begin its gradual transition to the red-giant stage.

The energy produced in the Sun in the manner described above is not uniformly emitted by its surface. As early as 600 B.C., without benefit of telescopes, Chinese observers had noted the phenomenon of sunspots, dark areas up to 130,000 kilometers in diameter.

Schematic representation of the twisting of
the lines of the Sun's magnetic field, caused
by the faster rotation of the Sun's lower
latitudes.

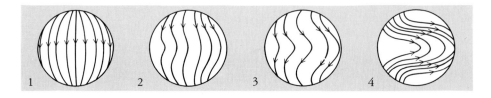

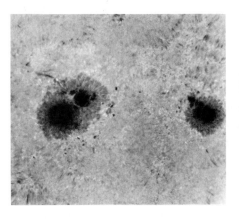

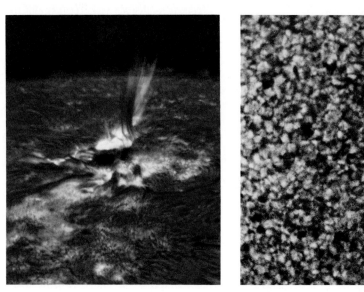

Above, sunspot pairs; at right, a portion of
the photosphere, a region of violent currents
and eruptions of gas; and, at far right,
granulation of the photosphere as seen
directly from above. The granules can be
interpreted as the tops of convection cells.

Sunspots are areas of the surface where the lines of the Sun's powerful internal magnetic fields break through the photosphere to form outside loops of strong local magnetic fields. These breakthroughs occur because the equatorial part of the Sun's surface rotates faster than other latitudes: equatorial rotation is about 25 days; half way between the poles and the equator, rotation is about 28 days. Thus the lines of the magnetic field that would normally run straight north and south between the poles of the Sun twist into dense east-west bundles and, at mid-latitudes, where the surface cannot always contain them, they pop out. As times goes on the sunspots occur at lower latitudes, so that a plot of their position against time produces a "butterfly" pattern.

The powerful magnetic fields present in the spots suppress the local upward heat flow from the deeper layers, so that the spots are some 1500 K cooler and

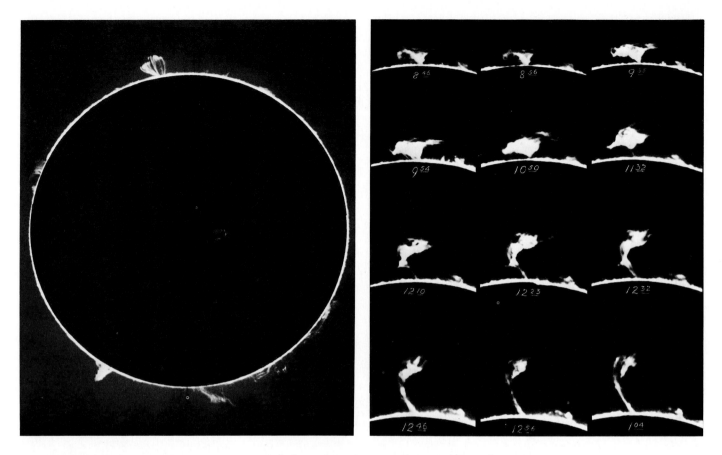

At left, huge looping prominences that can be seen at the edge of the solar disk when the rest of the Sun is obstructed by an opaque disk in the telescope. Such prominences can reach heights of hundreds of thousands of kilometers, far greater than the diameter of the Earth. At right is a time sequence showing the growth and motion of a prominence. The motion follows the lines of a loop of magnetic field protruding from sunspots.

so, darker, than the rest of the visible surface. The spots occur in pairs, one spot at each end of the loop of magnetic field, and therefore they have opposite magnetic polarity, one being a local north pole while its partner is a local south pole. In one hemisphere, say the southern, the leading spot of a pair—that is, the most eastward spot—might have a northern polarity, while the trailing spot would have southern polarity; in the other hemisphere, the order would be reversed. Every 11 or so years the spots essentially disappear and, when they reappear during the next period, show reversed polarity; thus, in our example, the most eastward spot in the northern hemisphere would have southern polarity while its partner had northern polarity.

The solar surface seen through various filters shows a complicated and rapidly varying structure much like the surface of molten iron or some heaving brew in a cauldron. What we see are granules, which are the tops of convection cells, in the centers of which the hot and, therefore, less-dense gases from the deeper layers rise, cool at the surface, thereby becoming denser again, and then sink along the granule boundaries. A particular granule, which is typically some 330 to 1300 kilometers across, ranging from about the width to the length of the state of California, exists perhaps ten minutes. Some correlation in behavior among neighboring granules reveals the existence, at greater depths, of larger units—called supergranules—of about 33,000-kilometer diameter.

Spicules, jets of hydrogen that rise about 10,000 kilometers and then rapidly dissipate. They appear darker than the rest of the surface because they are cooler.

The breakthrough of magnetic lines in sunspots is associated with an explosive local release of enormous amounts of energy that rapidly carries hot, electrically charged and luminous hydrogen gases along huge loops of magnetic field. These spectacular outbursts, or prominences, first seen when the solar disk was hidden by the Moon during solar eclipses, can be 200,000 kilometers high, with a gas velocity of about 1000 kilometers per second. Sometimes the normally dark gases, streaming above the sunspots, emit light, shining brightly when the atoms of the gases are excited by a burst of ultraviolet, X-ray, and other radiation from a sunspot. These bright gases are called flares. On a much smaller scale are the streaks of gases, called spicules, like bursting bubbles above the granules of the photosphere.

Attempts have been made to relate the apparent 11-year cycle of sunspots to cyclic phenomena on the Earth—from difficulties in radio reception to variation in sea level or length of the day, to outbreaks of revolution, and, even changes in the price of eggs. Of the few reasonably valid correlations of this kind, probably the best established stems from the fact that, although the fraction of the solar disk occupied by sunspots is exceedingly small, the magnetic perturbations associated with them can affect the upper layers of our atmosphere and alter, to a small degree, the heat reaching us from the Sun. Naturally this could affect the Earth's climate and thus, perhaps, crops and related matters.

Some confirmation of this may be found in the most famous perturbation of the sunspot cycle recorded—the so-called Maunder (after British astronomer E. Walter Maunder) minimum, an almost total absence of spots between 1640 and

Maunder's "butterfly" diagram, showing that the sunspots of each new cycle start near 40° latitude and, as the cycle progresses, appear at lower latitudes.

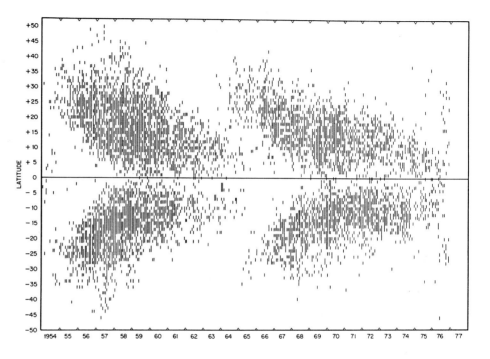

1715. Its occurrence correlates with historical records and tree ring evidence of unusually cold winters and heavy snow falls during the 75-year interval. The tree rings are narrower in cold winters and also show an unusually high concentration of radioactive carbon (^{14}C). This particular kind of carbon is produced high in the Earth's atmosphere by cosmic radiation coming from the galaxy. We know that during the so-called quiet Sun, when there are few sunspots and low solar wind, there is about ten times as much ^{14}C as during solar storms or an *active* Sun. This correlation is understandable in terms of the effect of the solar magnetic field and wind on the terrestrial magnetic field. It has been suggested that some ice ages were the result of prolonged periods of low sunspot activity.

Lying above the photosphere, the visible solar surface, is the solar atmosphere, which consists of the chromosphere and the corona. The chromosphere, about 8,000 kilometers thick, consists mostly of hydrogen emitting its characteristic reddish light. Although it is essentially transparent, the light coming through it is partly absorbed by various elements, giving a good indication of the solar composition close to the surface. The average temperature of the chromosphere is about 1500 K, although it is as high as 100,000 K in its outer region. As with the prominences, the chromosphere has to be observed either during solar eclipses or with special filters that suppress the light of the bright solar

The thin, luminous solar corona, seen when the light from the rest of the surface is obstructed by a disk. A coronal hole is a region of powerful X-ray and other high-energy emission that occurs where the corona is absent, usually near the poles.

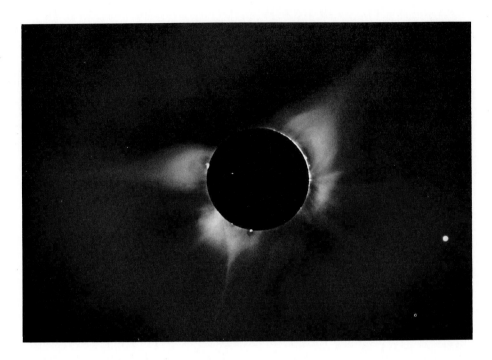

surface. The source of the high temperatures in the chromosphere is believed to be solar pulsations, which are described later.

The outermost layer of the Sun is the spectacular corona, first observed during solar eclipses. It is huge, its dimension comparable to that of the rest of the Sun itself. Studies indicate that it is made of very thin but hot—a few million degrees Kelvin—hydrogen gas in which the hydrogen atoms have been broken into hydrogen nuclei—protons (positively charged particles)—and electrons (negatively charged particles). These particles travel out into space at about 800 kilometers per second without colliding with each other. The flow of these particles constitutes the solar wind, in which about a million tons of ionized hydrogen leaves the Sun per second carrying with it the solar magnetic field. This "wind," which is, in effect, an extension of the corona, has been observed by spacecraft well beyond the orbit of Saturn, which is ten times farther from the Sun than the Earth. Solar wind varies considerably in intensity, but a typical flow at Earth's distance from the Sun is 100 million protons per square centimeter per second, which would be very dangerous for life on the Earth; fortunately our atmosphere prevents the solar wind from reaching the Earth's surface directly.

Temporary absences of a visible corona in many parts of the Sun were recently observed, especially near its polar regions. These coronal "holes" (which can persist for months) are regions that are colder (hence darker) than the usual visible corona, and are associated with very intense streams of the solar wind as well as with enormous outflows of radiation such as X-ray and ultraviolet, which are produced in the deeper layers of the Sun. These intense streams of the solar wind and the associated intense bursts of the solar magnetic field lines perturb severely the terrestrial magnetic field and the upper layers of our atmosphere, with resultant disruption of radio communication.

That the temperatures of the chromosphere and the corona are much higher than that of the photosphere may be accounted for by oscillations and pulsations of the Sun's surface of various frequencies and intensities. Those of one group pulse about once every 5 minutes and occur anywhere on the solar surface, although in any one area they last for only several pulses. Such an area, which may be tens of thousands of kilometers across and as much as several percent of the solar mass, will move about 10 miles up and down relative to the center. These pulsations reveal localized accordion-like compressions and extensions across the thickness of the outer 25 to 35 percent of the solar radius. This dimension agrees roughly with the thickness of that part of the Sun where heat transport occurs by convection.

The Sun also oscillates as a whole in a manner not unlike that of a large bell. The periods of these oscillations are as long as an hour or more and show as changes in the diameter of the sun of about one and one half to three kilometers. These small and slow oscillations are usually observed by measuring, very precisely, small shifts in the position of various lines in the solar spectrum, and there is a rather good agreement between the observations and the theoretical predictions based on what we know about the solar interior. These oscillations and pulsations of the solar surface are thought to account for the enormous differences in temperature between the Sun's surface and the solar atmosphere by producing shock waves, which travel outward to the thin outer layers of the solar atmosphere, where they become supersonic, exciting the gases in the outer layers to higher temperatures (and, consequently, to emission of light).

There is no other star about which we know as much as we do about our Sun. We understand quite well its internal and external structure and even its not necessarily predictable variability, which has many effects on us.

ESSAY II

THE FORMATION OF THE PLANETS AND THEIR SATELLITES

Although the basically simple Copernican structure of the solar system, with the Sun at the center and the planets revolving around it, seems obvious to us, its discovery and acceptance were very slow. Discovery was hampered by lack of good instruments, and acceptance by religious and philosophical dogma that asserted that Earth, being the place where Man, who was created by God, lived, must be the center of the universe. Even Copernicus, brilliant as he was, hesitated to go against the Greek insistence on ideally perfect circular orbits of planets, initially around Earth and then, in his own theory, around the Sun.

Our real understanding of the laws of planetary motions, which began toward the end of the sixteenth century, must be forever associated with Kepler, a German theorist who was assistant to Danish astronomer Tycho Brahe, and with Newton. It was Kepler who deduced by observation the geometric laws of planetary motions that were later derived theoretically from Newton's laws of gravitational attraction. However, Kepler could never have formulated his laws had he not had a vast amount (for the time) of highly precise observations on which to base them. These were made in Tycho Brahe's four observatories, which he built with his own funds and with assistance from Denmark's Frederick II.

The formation of stars, and of our Sun in particular, is rather well understood because there are billions of observable stars and we have been able to compare

Facing page: *Artists conception of a stage in the accretion of materials leading to the formation of the earth.*

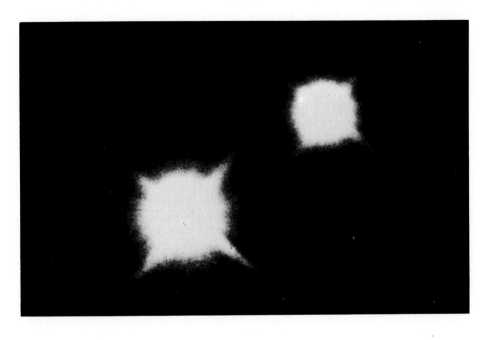

Alpha Centauri, second closest star to the Earth, is 4.3 light-years distant. It is actually a star system, composed of Alpha Centauri A (left) and Alpha Centauri B.

them, to use statistical methods to classify them, and to propose and test theories of their birth, life, and death. These theories do not require making questionable or *ad hoc* assumptions. This is not quite the case so far as the origin of our Sun's planets is concerned. If there were other planetary systems that we could observe at various stages of formation, we could compare their structures to that of the solar system and, perhaps, deduce and justify some general similarities among them and among the processes of formation. From observation and statistical analysis of various kinds of stars we conclude that there are billions of stars very similar to our Sun. All these stars could have acquired, during their formation, planetary systems of their own. Unfortunately, in spite of these excellent reasons to believe that our solar system is not unique, no planets outside our solar system have yet been observed. This is chiefly because, in contrast to stars, planets shine only by reflected light, so that, at the great distances involved, any outside our solar system would be too faint to be visible even with the most advanced telescopes. Furthermore, planets are likely to be much smaller than the stars around which they orbit, so that even the brightest might not be detectable in the glare of its sun.

There are many questions concerning the origin of the planets, chief among them being whether they formed from the same cloud as the Sun and what triggered their formation. There is evidence that the whole process took less than 100 million years some 4 to 5 billion years ago (as determined by radioactive dating of matter from meterorites). Also not clear is why there is a certain regularity in the distances of the planets from the Sun and why the inner, or terrestrial, planets (Mercury, Venus, Earth, and Mars) and the outer planets (Jupiter, Saturn, Uranus, and Neptune) differ so much in size and chemistry. It is thus not surprising that over the last few decades many theories of the formation of planets have been proposed, some quite ingenious. This situation is in striking contrast to speculation concerning the formation of stars, for which there is only one theory, which, with small variants, is generally accepted.

The variety of theories of the formation of planets reflects the difficulties of devising one when the observational data are meager. The most generally accepted theory of the origin of our planets is that, after the Sun was formed, whatever was left over from the primitive solar nebula became a disk rotating around the Sun, and the planets condensed out of this disk. Before describing this process, however, it is worth mentioning briefly two other schools of thought, which illustrate the variety of the points of view. One of them assumed that the planets were the results of perturbations of the Sun by a passing star; another, that the planets were formed from a galactic cloud other than the one that gave birth to the Sun. The first view is exemplified by the theory of M.M. Woolfson, of the University of York, who proposed that a star of some 100 solar masses passed near our Sun at close range and raised a tide on its surface, very much like the tides produced in the Earth's oceans by the Moon. The top of the tide was pulled from the Sun and eventually formed a planet. The loss of material by the Sun produced a huge wave on the solar surface that traveled around the Sun. When the wave, back at its starting point, faced the same passing star, its top, in turn, was pulled off. This new loss of matter created a new wave, and so on, until all nine planets and the asteroids were formed in regular sequence. The theory encounters several objections, one being that the likelihood that two stars would come so close to each other is very low. In a universe in which a typical star were the size of a ping-pong ball, the average distance between stars would be 650 kilometers. Another difficulty concerns the question of why the chunks of hot solar matter, which would have come from the outer, mostly rarefied, gaseous layers of the Sun, would collapse gravitationally to form individual planets rather than disperse or be pulled back by the Sun.

The other theory, first proposed by Swedish physicist Hannes Alfvén, now at the University of California, and later elaborated jointly with Gustav Arrhenius, from the same institution, suggests that after the Sun formed from an interstellar cloud it traveled through the galaxy and encountered another cloud of interstellar matter that was richer in heavy elements than the Sun itself. Part of this cloud was captured by the Sun, forming a nebula that revolved around the Sun and later broke up into individual planets. According to Alfvén, a major role in these processes was played by the Sun's magnetic field, which penetrates the whole solar system and can affect the motion of matter. This motion, combined with the effect of solar radiation, led to a partial separation of lighter elements from heavier and, at the same time, to the formation of a few chemically distinct streams of matter. These streams condensed later into planets. This sequence of events could account for the considerable difference of chemistry and density between the terrestrial planets, which are close to the Sun, and the giant planets, which are much farther from it. This theory, however, does not lead to results that can easily be compared and validated quantitatively by observation.

The most widely accepted theory is based on the nebular hypothesis proposed in one form in 1755 by German philosopher Immanuel Kant and, in 1796, in a different form, by French scientist Laplace.

The various stages of the formation of the Sun and the planets from a rotating nebula according to Laplace.

Kant suggested that the Sun was formed in the middle of a cloud and that, when it reached approximately its present size and became stable, there was still enough matter left over to form a nebula that revolved around it and from which the planets condensed. In Laplace's theory, on the other hand, the planets formed from material ejected into the nebula by the spinning Sun before it reached its stable form. In its modern version, this theory postulates that the initial, slowly rotating nebula was much bigger than the present planetary system but, under the influence of its own gravity, contracted, so that its rate of spin increased. (This contraction occurs because, for a body of a given mass, the product of the rate of spin and the diameter is a quantity that is conserved. Thus, reduction in diameter requires a compensating increase in rate of spin.) The density in the center was particularly high and led to formation of a protosun and so, eventually, the sun. The planets were formed from the remaining parts of the nebula. It follows that the planets and all other bodies in the solar system should have basically the same original composition as the Sun. This is, of course, not the case—there are, for example, significant differences between the densities of the inner and outer planets—and so, at first glance, the Laplace theory in one form or another seems inadequate. However, its recent refinements, discussed below, appear to resolve these puzzles very satisfactorily.

The evolution of the early nebula that revolved around the Sun has been partially elucidated by recent computer simula-

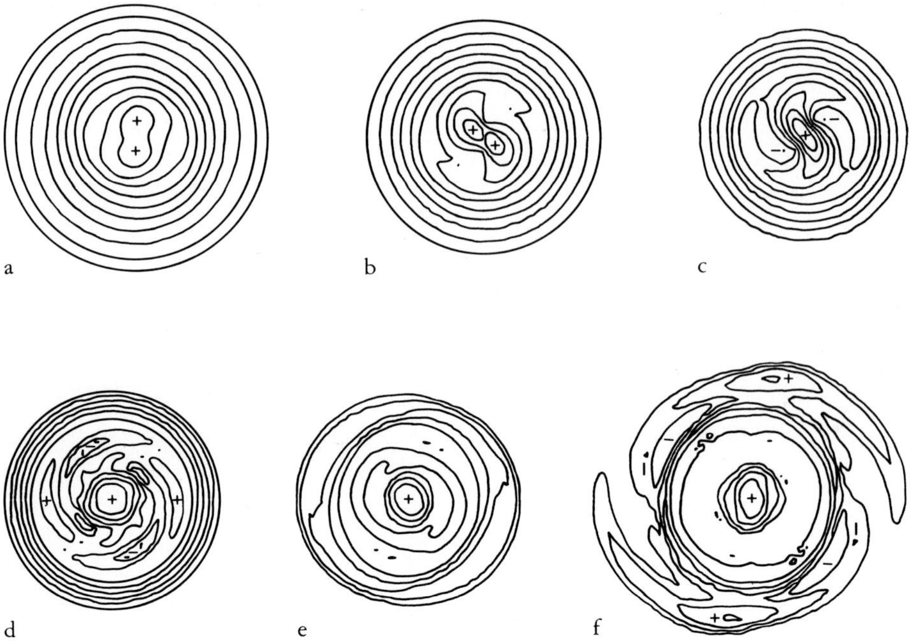

Computer simulation of the evolution of a rotating solar nebula. The central condensation becomes the Sun, and the arms, together with additional infalling matter, form a disk from which the planets eventually condense. Details of the process depend on initial conditions. The signs + and − indicate, respectively, high and low densities of matter.

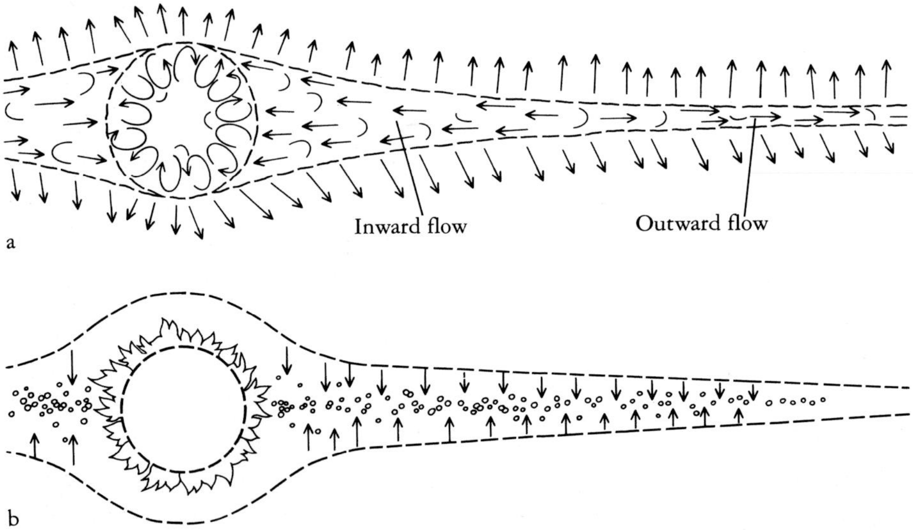

a   b   c

d   e   f

Formation of the accretion disk around the Sun: a indicates initial motion of gas, and b, subsequent motion of dust, which forms when the disk cools. Part of the initial gas flow is toward the protosun and part away from it (expanding the disk). Particles moving toward the mid-plane are gravitationally attracted.

Inward flow   Outward flow

a

b

tions of this process. It seems that, depending upon the starting conditions, there forms either a central condensation of gas with a ring or arms or an enormous, rather regular and fat disk, called an accretion disk. Studies in detail by British, American, and Soviet astrophysicists— Lynden-Bell, Pringle, Cameron, Safranov, and others—show that initially the disk is not completely detached from the rest of the gas cloud from which it and the Sun were formed, and thus it keeps growing in mass and size. The resulting compression raises the temperature of the disk and, when it becomes so high that material evaporates as fast as it is accreted, growth

The formation of the planets from dust particles according to Cameron. Particles accrete into clumps by collision (a). The clumps fall to the mid-plane of the disk (b), where they form planetesimals (c). The planetesimals accrete by gravitational attraction (d–f) into planets (g), which, if large enough, attract gases and thus acquire an atmosphere (h,i).

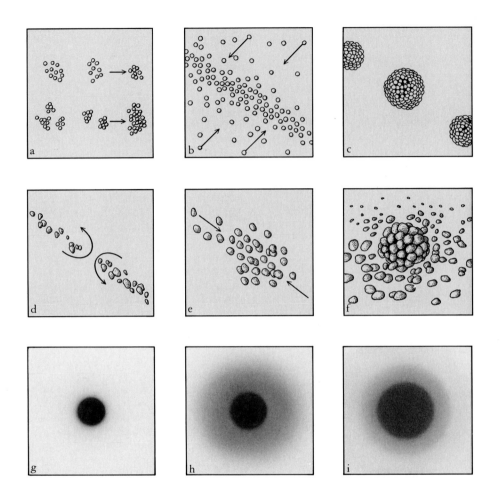

stops. From then on, that part of the nebula that is nearer to the Sun, about 20 percent of it, slowly falls into the Sun.

Further evolution of the remaining 80 percent of the disk resembles, in many ways, the familiar mechanism of rain formation, in which water vapor from the oceans rises and condenses into droplets in the cooler regions of the atmosphere. The initially small droplets are supported by air currents but, as they grow, they drift down, the bigger ones falling faster than the smaller. A bigger droplet encounters, during its fast descent, more other drop-

lets than do the slower, smaller ones, and so grows faster. If the temperature is low enough, the drops freeze and hail forms rather than rain. Calculations show that a similar process goes on in the gradually cooling nebular disk, with the difference that solid grains rather than droplets are formed and these drift not "down," but toward the mid-plane of the disk. These grains grow until some become so large that they begin to attract other grains gravitationally. Eventually the continuing accretion forms larger bodies, the planetesimals. These planetesimals, in turn, merge into still larger units called proto-planets, which consist primarily of solids. When these protoplanets cease to grow they become stable planets. It is natural that the first substances to aggregate into grains were solids such as rock and, later, ice. The lighter gases, hydrogen and helium, did not condense but were gravitationally attracted by the planets and formed atmospheres. It is believed that this is how the terrestrial planets were formed. The condensation of the outer planets (Jupiter, Saturn, and so on) was similar, except that in the outer parts of the nebula there was enough space and material to form local non-uniformities, which were big enough to lead to gravitational instabilities. These instabilities gradually produced enormous protoplanets made of gas and dust. It was only when these protoplanets were sufficiently large that the gradual cooling and contraction, and condensation into many planetesimals, occurred—by processes similar to those for the terrestrial planets. It has been shown that, nearer to the Sun than Jupiter, gaseous protoplanets could not have formed without first having passed through the intermediate planetesimal stage. Again, rocks and metals, which melt at high temperatures, were the first to aggregate, giving these planets huge rocky cores.

Unfortunately, because of various gravitational, mechanical, and chemical factors, a quantitative understanding of the rates of these various processes is still rather elusive. Part of the problem is that an encounter between two solid objects can lead, depending on their size and velocity, either to their breakup or to accretion. Experiments have shown that particles in low-velocity collisions, are more likely to stick together, especially if they contain some ice. Hard particles in high-velocity collision usually break up. For bodies heavy enough, gravity will keep together the pieces of partly shattered objects, so that accretion will follow. Whatever the details, it follows that all the planets started as bodies made of rocky substances that condense at rather high temperatures. Later they became large enough to attract and retain gaseous atmospheres. What was left of the solar nebula formed smaller nebulae revolving around each of the planets. One of the consequences of the uncertainties about the details of the processes here described is the difficulty of estimating the time necessary to form the planets. Actually, estimates vary from 100 thousand, to 100 million years. The latter figure seems unrealistically high.

The atmospheres of the outer planets are so big that the mean density (the ratio of mass to volume) of these enormous planets—that is, of the rocky cores *and* the atmospheres—is close to that of water; the inner planets, such as Earth, with relatively thin atmospheres or virtually none, have densities about five times that of water. As noted above, the differences between the densities of the terrestrial and the outer planets is often cited as an argument against the feature of the nebular hypothesis according to which all the planets formed from the same nebula. The following, based on recent developments, describes several ways to account for these differences. One is based on the fact that, between the Sun and a planet (just as between Earth and the Moon), there is always a critical point at which the gravitational attractions of the two bodies is equal and opposite. According to A.G.W. Cameron, of Harvard University, initially, these points lay outside the protoplanets, but, as the solar mass increased, the huge protoplanets came nearer and nearer to the Sun, and the critical points came to be located within their atmospheres. As a result, the outer layers of the protoplanets were stripped away

The point L, *where gravitational attraction between the Sun and a protoplanet are equal but opposite, can fall inside or outside the protoplanet. When it is inside, as in* b, *the protoplanet's outer layers (shaded portion) are stripped away and fall into the Sun.*

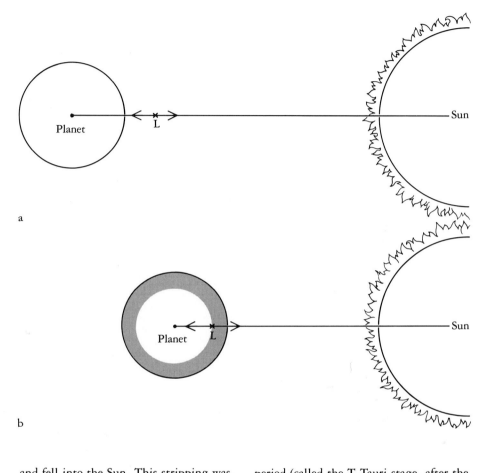

and fell into the Sun. This stripping was, of course, particularly effective for the inner planets, whose gases were at the same time being boiled off by the Sun's heat, which raised the planets' temperatures in a range from 1400 K (Mercury) to 500 K (Mars). The outer planets did not come near enough to the Sun to lose their enormous atmospheres in this way, and thus, their mean densities remain much lower.

The differences in density of the planets could also be understood as the result of an instability and flare-up that is believed to occur shortly after a new star, such as our Sun, is formed. During this period (called the T-Tauri stage, after the star in which it was first observed) and lasting as long as 100,000 years, a star's luminosity increases 30–40 times and a substantial amount of its outer layers is blown off. This event is the last fling of a young star before it settles down to its long and rather uneventful life. The solar T-Tauri wind was enormous, blowing away most, if not all, of the gaseous components of the remnants of the nebulae and of the early atmospheres surrounding the planets close to it. The intensity of the wind would decrease rapidly with distance from the Sun, so that when it reached the distant outer planets, it was too weak to

A T-Tauri star (at the center), one that is in its infancy and so still contracting and growing hotter. These stars can show great increases in brightness over relatively short periods. Note the streams of matter being ejected from the star.

have much influence, thus leaving the inner planets denser than the outer. Quite likely both the heating and the T-Tauri wind were present and account not only for the large differences between the inner and the outer planets but also for the small differences between the overall planetary composition and the solar composition. It should be noted that the concept of a T-Tauri stage in the development of our Sun and the associated wind has recently been criticized.

Just as a stream may move silt particles and sand grains but not pebbles, the solar T-Tauri wind, which consisted mostly of hydrogen, blew away the gases but could not remove the larger solid grains already formed in the nebulae surrounding the planets and revolving around them. These grains kept growing by accretion and coalescence until they formed larger masses and eventually became the satellites of the planets. At that time, temperatures in the nebulae around the planets were so low that ice grains as well as rock grains were present. Thus, it is quite understandable

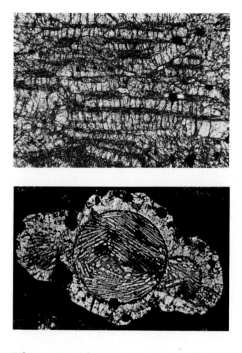

Thin sections of meteorites showing evidence of collision-induced shock (above) and melting and cooling. At these enlargements a quarter inch is equivalent to about 50 microns.

that many planetary satellites are either rocky or icy or, more usually, a mixture of rock and ice. Formation of the satellites of the outer planets was, therefore, a repetition on a small scale of the processes that led to the formation of the planets themselves. Accretion and growth of grains were undoubtedly the main mechanisms of formation of the satellites.

Similar processes are believed—although with much less certainty—to account for the existence of comets, which seem to be clusters of ices and dust coming to us from the outer reaches of the solar system. According to Cameron, at the end of the infall of matter, when the solar nebula in its early history reached its largest radius—about 600 astronomical units (or A.U., each unit the distance between the Sun and the Earth)—its outer regions cooled rapidly so that water vapor and other gases condensed, forming grains of various ices and other solids. Clusters of these solids, a few kilometers in diameter, were then formed and still exist, revolving very slowly around the whole solar system. Mutual collisions and perturbations, including the influence of passing stars, may have stretched this cloud 100-fold and spread it into a spherical shell around the whole planetary system. This is the so-called Oort (after Dutch astronomer Jan Oort) cloud. Another possibility is that the Oort cloud is a collection of icy clusters originally condensed within the planetary system (near Uranus or Neptune) that were subsequently thrown out by perturbations by the giant planets.

What is the likelihood that the planets and their satellites were formed as just described, in the modern version of the Kant-Laplace theory? Of great importance in this respect is the recent finding by R.I. Thompson, of the University of Arizona, and colleagues of at least one, if not two, objects in the sky that can best be described as very bright stars surrounded by accretion disks. These objects appear to provide the very important observational confirmation of the reality of preplanetary accretion disks. Such disks are not observed more frequently or easily because they are transient and are luminous for only relatively short periods. Probably because gravitational attraction between two small bodies is extremely difficult to observe, there has been no direct laboratory evidence for the occurrence of the theoretically predicted clustering—either on a large or on a small scale. On the other hand, there is plenty of experimental evidence for and good understanding of accretion of grains and growth of small droplets in vapors under all kinds of circumstances. As noted earlier, the quantitative aspects of growth and accretion through collision are not well understood.

As for the later stages of planetary growth, there is good evidence that violent collisions did occur between planetesimals and, at a later stage, between planetesimals and planets. Many of these collisions led to the formation of rocky fragments of all sizes and shapes, which strike planets and satellites. When they enter Earth's atmosphere, friction makes them white hot and they become visible as meteors. Those that do reach Earth's surface (called then, meteor*ites*) provide excellent information about the interplanetary processes. For instance, some meteorites show structural features that can be explained only as the result of enormous shocks or of slow cooling following melting induced by collisions. Similarly, all observed solid surfaces of planets and of satellites have huge craters that must have been produced by impacts of massive bodies of sizes one would expect to find among planetesimals. It is noteworthy that, as discussed in the Afterword, some meteorites found on Earth also contain biological matter of unknown origin.

Attempts have been made to clarify the

The density, at the various planetary orbits, of planetary matter were it distributed uniformly around the Sun. Although the terrestrial planets are smaller than the outer planets, the distances between their orbits are also smaller, so that their distributed density is greater. For asteroids this density is about 1000 times lower than it would be if density in the accretion disk decreased more or less uniformly with distance from the Sun.

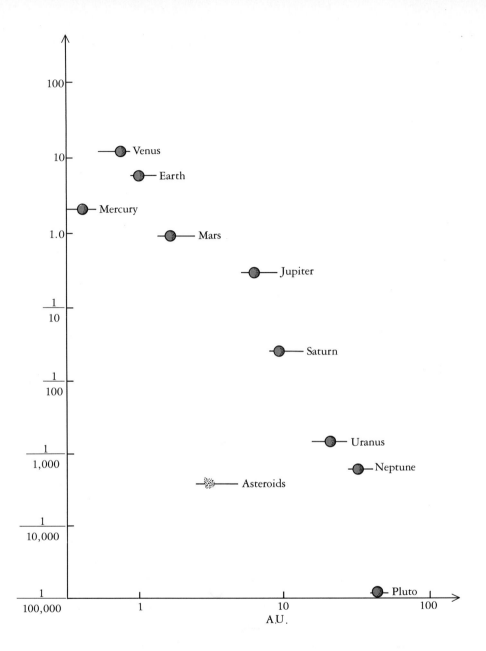

still considerable controversy concerning the formation of the planets by comparing the planetary system with systems of satellites. Jupiter, Saturn, and Uranus have, of course, many satellites and so one could, in principle, make comparisons among these three systems and draw conclusions about the formation of the planets. A difficulty stems from the fact that most of the outer satellites of these planets appear to be captured bodies, formed elsewhere, and not satellites that grew from the original nebulae around the planets. Furthermore, it is only very recently (from spacecraft missions) that information regarding their interiors has become available. The Moon and the two satellites of Mars are excluded from this comparison because their origin is still shrouded in mystery.

The origins of planets, satellites, comets, and meteorites seems to be thus fairly well understood. A problem still not completely resolved, however, is the mechanism that determined the location of planetary orbits around the Sun and the location of the orbits of satellites around the planets. At the end of the eighteenth century, a seeming regularity was noted in the spacing of the planets. If the distance between the Earth and the Sun is taken as unity, then the distances from the Sun of all the then known planets are as follows: Mercury, 0.39; Venus, 0.72; Earth, 1.0; Mars, 1.5; Jupiter, 5.2; Saturn, 9.5; and Uranus, 19.2. The German astronomers Titius, and later, Bode, pointed out that these numbers can be expressed reasonably well by the formula $(x + 4)/10$, where x is a number in the series 0, 3, 6, 12, 24, etc. in which, each number (except 3) is twice its predecessor. This Titius-Bode rule is not perfectly applicable: When Neptune and Pluto were discovered, it was found that Neptune does not follow the rule, while Pluto, does, although its origin is probably different from that of the other planets. Interestingly enough, for $x = 4$ one obtains the orbit of Vesta, one of the biggest asteroids. Other forms of the Titius-Bode rule have been generalized to apply also to the spacing of the inner satellites in the systems of Jupiter, Saturn, and Uranus. There have been many ingenious attempts to explain these apparent regularities. A recent one, made by Michael Torbett of the University of Texas, Richard Greenberg of the Planetary Science Institute, and the author, shows that the formation of the planets can be accounted for by considering the influence of Jupiter on the early evolution of the solar nebula. Jupiter, being by far the biggest of all planets, was undoubtedly formed first, and its huge gravity must have affected the process of formation of the other planets by perturbing, in a regular and predictable way, the subsequent evolution of the solar nebula and, in particular, the motions of and collisions between planetesimals. These, in turn, led to the preferential formation of planets at particular distances from the Sun. The positions of the planets calculated in this manner agree with observation as well as with the Titius-Bode rule.

Where the Titius-Bode rule predicts a major planet between Mars and Jupiter, the boundary between the inner and the outer planets, there is instead a belt of thousands of asteroids. Many attempts have been made to explain this gap. One, by Michael Torbett and the author, points out that Jupiter and the other huge outer planets had already formed at a time when Mars and the other inner planets were still growing by the slow process of accretion. Thus, at that time, there was in the region between Mars and Jupiter still much solid matter, ranging from small grains to planetesimals orbiting the Sun. The gravity of the gigantic Jupiter would have perturbed the orbits of the smaller bodies and led to increased collisions among them, and consequently to breakup and even ejection of material. At the time, when the Sun's mass was still growing, it was gradually attracting everything more strongly, so that the radii of the orbit of Jupiter and of the orbits of the planetesimals would have slowly decreased. During this process, the perturbing influence of Jupiter spread over an ever wider range of distances between Jupiter and Mars, and, as a result, much of the matter of the disk between these two planets was broken up and ejected. The striking absence, then, of a major planet, the low density of matter in the asteroidal belt, and the presence of gaps in this belt can be accounted for in this manner.

Though we have so far treated our solar system as a more or less isolated entity uninfluenced by other components and events in the galaxy, this is not entirely the case. We know that all parts of the galaxy—the stars, the dust clouds, and all other matter—are in continuous motion with respect to each other and, in particular, that the Sun, with its planetary system, travels at about 20 kilometers per second toward the star Vega. Therefore, there is a good chance that on its way it has encountered and will continue to encounter, various interstellar clouds. It is estimated that, since the Sun was formed, about 120 such encounters have taken place, with capture by the Sun of matter equal to about one ten-thousandth of its present mass. What, if any, effect on the planets these encounters had or will have is not known, although there have been speculations about correlations with various drastic events on Earth such as the rapid and unexplained end of the age of reptiles.

Other intragalactic events were of great importance for the chemistry of our solar system. One is the deaths of stars that occurred before our Sun was born. A supernova explosion occurs when an old star, perhaps four or more times more massive than our Sun, collapses within a few minutes, releasing in this short time energy comparable to that emitted by our Sun in one billion years—an event of staggering magnitude. The enormous blast of energy blows off most of the star to form a shell of gas expanding at a rate of tens of thousands of kilometers per second. The remnant of this event is an extremely dense body called a neutron star, having a diam-

Development of a nova (Nova Cygni 1975) over a period of about two days. Arrows indicate position of star (virtually invisible) before the explosion.

eter of only about 10 kilometers. A neutron star forms so rapidly that the magnetic field of the initial star does not have time to escape, and, squeezed into the volume of the new star, reaches enormous intensities. Such a high-energy field leads to the production of a beam of intense radiation that we can observe. Since the neutron star rotates rapidly, the beam sweeps through space like the beam from a lighthouse, giving the appearance to observers on the Earth of rapid pulses of radiation, which is why these stars were originally called pulsars. The impact of their discovery (by Anthony Hewish at Cambridge University) was enormous, and for a time gave rise to speculation that the pulses might be signals from other civilizations. Actually the pulses were extremely regular, though slowing almost imperceptibly, and contained no message or code. Thus the more prosaic but scientifically sounder suggestion by Thomas Gold, at Cornell University, that they are rotating neutron stars, became accepted.

The gravitational forces of a neutron star or of a white dwarf that has a companion star circling it closely, may be strong enough to strip the outer layers of the companion. These layers, falling rapidly onto the small, dense object, produce so much heat that a nuclear reaction may start that leads to a perhaps 10,000-fold brightening of the old star. This is a nova.

It is believed that remnants of supernova events in even more massive stars are the mysterious black holes, which are of fundamental interest because they permit comparing various predictions of the general relativity theory with observations. Black holes got their name—suggested by John A. Wheeler, now at the University of Texas—from the fact that they are so massive and have such high gravity that not even light can escape from them, so that they are thus not directly observable. This peculiar property is another well-confirmed consequence of the theory of relativity, according to which, gravity affects the path of light in a way similar to the way it affects matter, say a thrown stone—both fall down. It is only the black hole, however, that has a gravity that is high enough to prevent the escape of light.

These events are of enormous importance for us as described below. The sequence of nuclear reactions, mentioned above, that follows the formation of helium from hydrogen leads to the formation of other elements, but stops at iron—making it the heaviest element that can be formed in this manner in the center of large stars. Where then do the heavier elements come from? It appears that the formation of these elements requires much more energetic collisions between atomic particles than are obtainable even in the hottest stellar interiors, and it is in the huge and rare supernova explosions, discussed above, that these conditions are found. This difference between the origin of elements no heavier than iron and of those that are heavier suggests that there should be far more of the light elements than of the heavy ones, and in fact the abundance of the former in the cosmos is some 10,000 to 100,000 times that of the latter. As time goes on, more supernovas take place, and each adds its share of the freshly formed heavier elements to interstellar matter. Although

hydrogen is still by far the most abundant element, younger stars such as our Sun contain more heavier elements than do the older ones.*

It is, then, a remarkable fact that all the lighter elements on Earth and the other planets, from which we ourselves are made and which we know and use—carbon, aluminum, oxygen, sulfur, zinc, silicon—were made in nuclear ovens in the centers of stars that disappeared billions of years ago, and that all the heavier elements, such as gold and platinum, were made in equally ancient, rare supernova explosions. It is clear how vain was the ancient dream of alchemists of transmuting base metals into gold by purely chemical means.

We have seen that the origin and the general structure of the solar system, as well as its motion, were fairly well understood for some time. The past decade or two has brought about an incredible advance, if not revolution, in our knowledge of the planets themselves and their satellites. This progress has its roots in the spectacular results of spacecraft missions to most of the planets and in greatly improved observational instruments.

---

*These differences in the chemical composition of stars of various ages are easily observed by spectrographic analysis, in which light from a star is split by a prism into the various colors characteristic of the materials of which it is composed.

CHAPTER 2   THE HOTTEST PLANETS:
MERCURY AND VENUS

CHAPTER 2

THE HOTTEST PLANETS: MERCURY AND VENUS

It is not surprising that conditions on Mercury and Venus, the only planets nearer to the Sun than the Earth, are rather extreme and that there was no—and probably never will be any—life on them: Mercury has virtually no atmosphere, and that of Venus is so dense and so hot that water in its liquid form could not exist. Though very different in some ways, these two planets are similar in others. They are, for example, the only planets without satellites—which is usually interpreted as the result of their proximity to the Sun, with which they could not compete in capturing a passing body or in holding excess nebular material.

Before the age of spacecraft and before the development of powerful radar beams, both planets were essentially inaccessible for detailed study from the Earth: Mercury because it is so close to the bright Sun, and Venus because of its solid cloud cover, which explains its unusual brightness. For this reason, the present richness of information, often quite startling, is of recent origin and much of it incomplete.

MERCURY

Venus, Mars, Jupiter, and Saturn, the brighter planets, were discovered early because an unaided but patient eye could notice their fairly rapid motion with respect to the stars, which appear to be fixed in the firmament. Among the

Preceding page: *False-color picture of cloud-shrouded Venus. The large-scale patterns are exceptionally stable.*

Mercury and Venus, because they are so close to the Sun and inside our orbit are always seen near sunset or sunrise and show phases, much as the Moon does. The angle of the orbits as seen from the Earth is actually much smaller than shown here for illustrative purposes, so that an observer on Earth sees them almost edge-on.

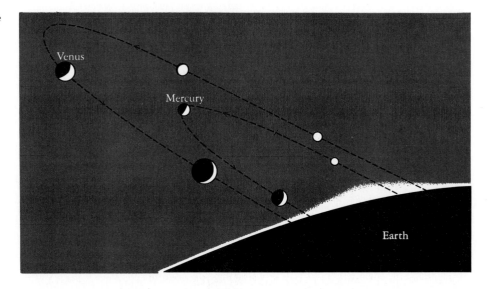

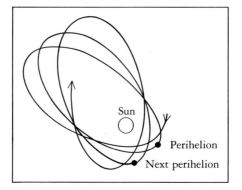

The point in the orbit of Mercury at which it is nearest the Sun (perihelion) moves very slowly in the same direction as the planet itself, a motion that played an important role in one of the observational confirmations of the validity of Einstein's theory of relativity. The degree of movement of the perihelion shown here is greatly exaggerated.

dimmer planets, Mercury was difficult to observe even after the first telescopes were constructed, in part because it is small and in part because it is so close to the glare of the Sun that, under the best circumstances, it could be seen for less than two hours immediately after sunset or before sunrise, when the sky is quite bright. In fact it has been said that Copernicus, founder of the present concept of our planetary system—with the Sun, rather than the Earth as its center—never saw Mercury, although he knew of its existence.

Mercury's orbit around the Sun is rather unusual because it is less nearly circular—that is, more elliptical—than the orbits of other planets and because it deviates from the ecliptic, the plane of the Earth's orbit more than most other planets. Historically, the precession of Mercury's orbit—that is, a continuous, gradual shift in the direction of its longer axis—played an important role in the early confirmation of Einstein's theory of relativity. According to the theory, the gravity of one Sun,—and of any other mass—deforms its surrounding space so that light passing near the Sun does not travel in a straight line, but is slightly bent. This deformation of space also affects the motion of bodies around the Sun. Prediction of this shift in Mercury's orbit by traditional Newtonian mechanics did not agree with observation, while that based on Einstein's equations did.

As the planet nearest the Sun, Mercury has the shortest period of revolution around it—only 88 days—and the highest velocity, which makes its name,

A photomosaic of Mercury composed of images obtained by the Mariner 10 spacecraft. Resolution—the size of objects that can be discerned—is about 2 kilometers.

Mercury was the winged messenger of the gods in Roman mythology particularly apt. Mercury's rotation is about 59 days. Usually there is, as for the Earth, no correlation between a planet's rotation and revolution, but curiously enough, the ratio of Mercury's revolution and rotation periods is almost exactly 3 to 2. An explanation of this ratio may be that the planet has two opposing slight equatorial bulges, each of which alternately faces the Sun at its closest approach to it. As a result of this simple 3:2 relationship and of the great orbital eccentricity, Mercury's equator is not uniformly heated by the Sun but receives two

The Caloris basin on Mercury, produced by the impact of a body at least 160 kilometers in diameter; the diameter of the basin itself is 1600 kilometers. A ring of mountains up to 2 kilometers high forms the outer edge of the basin, and compression cracks are clearly visible. In this view about a third of the ring is visible, curving from roughly the center to the upper and lower left-hand corners of the picture.

and a half times more heat at longitudes 0° and 180° than at longitudes 90° and 270°. Mercury also receives six times as much heat as the Earth from the Sun, which appears in its sky about 2 and 3 times as big (depending on Mercury's position in orbit) as it does in our sky. There is little surprise, then, that the highest temperature on Mercury, which has no atmosphere, is 700 K (430° C), or more than seven times the lowest temperatures, 93 K (−180° C). On the Earth, such differences are much smaller because of the mitigating effects of oceanic and atmospheric currents and of the large volumes of water and air present in the oceans and the atmosphere.

Until the American spacecraft Mariner 10 flew by Mercury in 1974, our knowledge of the planet was minimal because of the brightness of the nearby Sun, as mentioned earlier. Mariner 10 was so directed that it went into orbit around the Sun, encountering Mercury three times before it ceased to function. It provided us with thousands of images and excellent, often startling data.

Results confirmed the expectations expressed by American selenologist Ralph B. Baldwin, in 1949, that the surface of Mercury is heavily cratered and resembles that of our moon. This cratering on Mercury suggests that, when various bodies bombarded the Moon, probably a similar swarm of interplanetary matter bombarded Mercury. The similarity in appearance of the surfaces of the two bodies may indicate also a chemical similarity. Very likely, at the time that Mercury and the Moon were bombarded the same was happening to the Earth. The Earth's surface, however, was partly protected by its atmosphere, while erosion of its surface and geological activity obliterated most of the craters that *were* formed. The 1980 Voyager flight showed that a similar bombardment occurred in the outer parts of the solar system. We have, thus, strong evidence of the presence of a swarm of bodies of all sizes—from small meteorites up to bodies perhaps 1500 kilometers in diameter—in the solar system at an epoch that, according to radioactive dating of lunar rocks, can be placed at some 4 billion years ago. The picture agrees with our intuitive ideas about the formation of planets by accretion and coalescence of planetesimals and with the notion that, once the planets were formed, there were still enough bits and pieces of matter left in the solar system to produce significant cratering of all planetary surfaces.

Evidence of a tremendous impact on Mercury's surface by a body 160 kilometers or more in diameter is the Caloris basin, with a diameter of 1600 kilometers. The surface shows not only a series of huge concentric rings and cracks, but also, on the antipode of Caloris—that is, where a line from Caloris through Mercury's center would emerge—we see a surface that is heavily deformed in a manner resembling no other area of the planet. It is almost certain that the powerful shock of the Caloris impact traveled not only through the planet, but also along its surface, becoming focused on the antipode and producing the peculiar surface

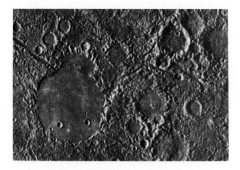

The antipode of Caloris, focus of the shock produced by the impact on the opposite side of the planet.

configuration seen there. The only other possible examples on a planetary scale of such effects are the antipodes of certain huge lunar craters and, perhaps, those on the Martian satellite Phobos—although other explanations for these features have been put forward. Similar phenomena have been observed only in laboratory experiments on objects bombarded with high-speed projectiles.

Also noteworthy is the pattern of compression cracks in Mercury's rocky crust. This pattern of cracks agrees with the idea that the initial accretion in all the planets produced in their interiors much heat, which then gradually escaped, the smaller planets cooling more rapidly than the larger. Mercury's original temperature was so high that its interior was and perhaps is still liquid, permitting iron and other heavy compounds to drift to the center, forming a Moon-size core and leaving a layer of lighter rocks on the surface. Upon partial solidification, the planet shrank rapidly by 3 to 4 kilometers, producing the observed compression cracks in the crust. The relation between the cracks and the craters suggests that the shrinkage occurred after the end of the period of intense bombardment.

The biggest surprise was Mariner's discovery that Mercury has a magnetic field, though a very weak one. In order to possess a field, a planet must not only be rotating fairly rapidly, but also must have an interior that is hot, liquid, and electrically conducting. For these reasons, a magnetic field is a very sensitive indicator of the past and present conditions in a planet's interior. The high density of Mercury—about twice that of typical rocks—indicates that up to three-quarters of the diameter of the planet is made of iron and iron compounds, but because of its high temperature these components cannot be kept in a magnetized state. Yet the magnetic field measured by Mariner is too strong to be explained by magnetization of the cooler outer rocky crust of the planet. Unless these arguments are wrong, it seems necessary to conclude that, as on the Earth, the planet's deep interior is still liquid—kept so by heat from the decay of radioactive elements—and that a magnetic field is thus generated in spite of the planet's rather slow rate of rotation. A quantitative analysis of this conclusion gives important information about the evolution of the solar system because it permits evaluation of the heat of accretion, the rate of its loss, and the rate of present radioactive heating. The fact that the mean density of small Mercury is nearly the same as that of much bigger Venus and the Earth implies that it has relatively more iron and less rock. This result points to the very important role in planetary evolution of solar heating, which made it more difficult for the more volatile constituents of the nearer portions of the nebula to condense.

Whether a planet has a substantial atmosphere or not depends on the velocity of its gaseous molecules and atoms, which, in turn, depends on temperature. If the temperature, and therefore the velocity, is high they can escape the planet's

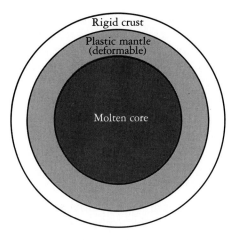

Mercury's interior, most of which is iron and iron compounds, partly molten in the core.

gravity, and thus, a massive planet can hold its atmosphere if it is not very hot; a small and hot planet would lose its atmosphere quickly. It was thus not surprising that Mariner observations confirmed the almost total absence of atmosphere on the small and hot Mercury. Whatever gas there is on Mercury is mainly a small amount of helium that either escapes from radioactive rocks or is continuously replenished by the intense solar wind. If Mercury had an early primitive hydrogen and helium atmosphere, it was certainly blown off during the solar T-Tauri stage. There is no evidence of the presence on Mercury of water either now or in the past. It seems that the heating and degassing of Mercury in its early history was very thorough and that no gases were left in its interior or in the crust to produce a secondary atmosphere at a later stage. There is very little doubt that there never was and never will be any life on Mercury.

VENUS

Over and above the early recognition of Venus as a very bright planet and its astrological and mythological impact on mankind's imagination, Venus played an important role in convincing Galileo that the planets revolve around the Sun—contrary to the then prevailing belief that the Earth was the center of the universe and all things celestial revolved around it. The argument is, for us, now very simple and irrefutable: during the course of its 225-day orbit around the Sun, Venus, being one quarter nearer to the Sun than the Earth, is sometimes between us and the Sun, sometimes—from our viewpoint—back of the Sun, and the rest of the time west or east of it. As a result, Venus shows phases, just as our Moon does, and we sometimes see its disk fully illuminated, sometimes mostly dark and, in between, as crescents. Galileo also noted that when Venus was seen fully illuminated it was small and far away, while at the time the crescents were narrowest, Venus appeared about seven times larger and thus nearer to us. Actually, to a terrestrial observer, Venus is brightest when it is close to being its farthest from the Sun in the sky. The only way to account for all these observations is to assume that Venus revolves around the Sun, much as our moon revolves around the Earth. Galileo rightly extended this conclusion to all planets and nearly lost his life for this heresy.

As soon as the distance to Venus and, thus, its actual mass and size could be estimated, it became clear that in these respects it was very similar to the Earth. We now know that its diameter is only 5 percent less than Earth's, and its mass 8 percent less. Naturally, there was much early speculation that its atmosphere and climate are similar to ours and that there must be life on it, a romantic idea ultimately dispelled but persisting in various forms into the 1960s.

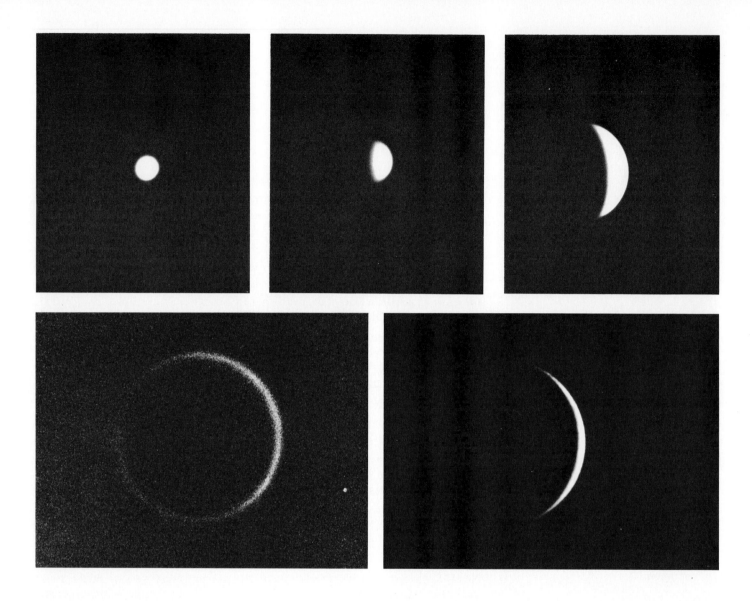

The phases (and more than five-fold differences in apparent size) of Venus, a phenomenon on which Galileo based his conviction that the planets rotate around the Sun and not the Earth.

The Soviet 1967, 1970, and 1975 Venera spacecraft, the probes that landed on the planet's surface, the American 1962 Mariner flyby, and the 1978 Pioneer orbiter, with its 5 probes, all provided excellent data about Venus' chemical composition and temperature variation with altitude. On its approach to the planet, the Mariner spacecraft made a sequence of photographs of the opaque deck of clouds, which lies at an altitude of 65 to 80 kilometers above the surface. The topmost layer of visible clouds moves around the planet in only 4 days, as compared to the 224-day period of rotation of the solid body of the planet. This difference, indicating persistent high-altitude winds of 1000 kilometers per hour, has been confirmed by direct measurements made with probes that descended to the surface, where winds reach speeds of only about 3 kilometers per hour.

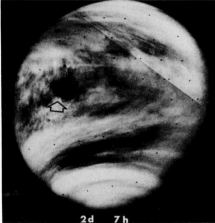

Movement of clouds on Venus. These views were taken at 7-hour intervals, indicating a rotation period of about 4 days. The arrow indicates the same cloud feature at the different times.

Studies made from the Earth or from spacecraft, with radar, which penetrates Venus' cloud layer, show not only that the planet rotates very slowly, but also that the direction of this motion is retrograde—this is, that it rotates in a direction opposite to that of most other planets, so that the Sun rises in the west and sets in the east. The period of rotation, 224 Earth days, is the slowest of all the planets and is longer than the period of the planet's revolution, so that Venus' day is about ten percent longer than its year. Interestingly enough, whenever Venus is closest to Earth, when both are on the same side of the Sun, it shows us approximately the same face. Originally, this regularity was taken to indicate the existence of some sort of an equatorial bulge not unlike those on Mercury, discussed above. Recent, more accurate data show, however, that the coincidence is not perfect and that there is an error of about 4 hours per period of rotation. It was thought that these interesting relationships might be related to tides raised by the Sun and by the Earth in Venus' solid body and in its very dense atmosphere. These effects turn out to be difficult to evaluate, however, and a generally accepted explanation of the retrograde revolution and its apparent relation to the Earth is not yet available.

The probes have shown also that atmospheric pressure at the surface is 90 times that on the Earth—nearly four times the highest pressure tolerated by a deep-sea diver. The mean temperature is about 750° K (480° C), high enough to melt lead and zinc. Thus, the earlier romantic speculation that the surface of the planet is livable, in our terrestrial sense, is wrong. In fact, no probe that landed on the surface operated longer than about one hour before being silenced by the heat and, perhaps, pressure. The two Soviet landers sent by Venera 9 and 10 spacecraft, which landed about 2000 kilometers apart, and the recent Venera 13 and 14 were able, however, to transmit a few photographs of the surface showing rocks a fraction of a meter in size. A simultaneous measurement of their radioactivity indicated that, on some sites, they resemble our basalts, and, on others, appear more like granite, of which many terrestrial mountain ranges, such as our Tetons, are composed. Basalt is of volcanic origin, while granite is evidence for geologic activity involving deformations and pressure-induced transformations in the crusts of continents. Most of the rocks seen in the Venera pictures have smooth and rounded edges, suggesting strong erosion (and corrosion) although, on one site, they are sharp, indicating that here they are probably of recent volcanic origin.

The main reason for Venus' extremely high surface temperature is the greenhouse effect, so called because of the ability of a greenhouse to trap radiant energy. The glass roof of a greenhouse is transparent to the sun's visible rays, so that a large fraction of them can enter and heat the soil and plants. These, in turn, reradiate heat, but this thermal radiation is of a wavelength, different from

The surface of Venus. The rocks visible range down to fractions of a meter. The toothed ring is a part of the Venera lander.

that of the incoming visible radiation, that cannot escape through the glass roof. The roof also prevents the escape by convection of heated air. Thus, the energy of the sunlight is trapped in the greenhouse and increases the inside temperature. On Venus, carbon dioxide and water vapor in the dense atmosphere and the opaque cloud cover, which incidentally, reflects back into space about 80 per cent of the solar light falling on it, play the role of the greenhouse roof. Venus' strong gravity prevents the hot atmosphere from rising much; on the other hand, the winds distribute the heat so that there is only a small difference between daytime and nighttime temperatures in spite of Venus' 8-month-long day.

Why does Venus have so thick an atmosphere and how did it originate? Since the primary (primordial) atmosphere of Venus and of the other terrestrial planets was lost when the solar T-Tauri wind blew them off soon after the planets were formed, the present, secondary, atmospheres are of different origin. Undoubtedly they are the result of volcanism and of the subsequent heating by stored internal heat and solar radiation. Besides lava and rock, volcanos emit large amounts of gases, such as carbon dioxide, water vapor, nitrogen, and sulfur compounds, depending on the chemical composition of the planets' liquid interiors. Laboratory studies show that heating of rock may lead to the release of gases that were trapped in small cavities when the rock solidified or that are a product of the rock's decomposition. This process of increased production of gas with increasing temperature accounts for the strength of the runaway greenhouse effect on Venus: the higher the temperature, the greater the volume of gas produced, and, consequently, the less transparent the planet's atmosphere and cloud deck; this leads to increased trapping of solar radiation, still higher temperature, and so on. Ultimately, the process reaches equilibrium.

Apart from a very small amount of "noble" gases, such as neon and argon, a few percent of nitrogen, about .01 percent of water, and small amounts of other gases, Venus' atmosphere is made of carbon dioxide, a gas we exhale and a typical product of the heating of certain rocks. The noble gases are so called because essentially they do not form chemical compounds with other elements and, as a result do not take part in transformations and various reactions, so that their presence or absence is a good indication of what happened during the first stages of the formation of the solar system. Among other gases is a carbon-oxygen-sulfur compound that forms near the surface. When carried to high altitudes, this gas is decomposed by solar radiation. The escaping sulfur interacts with traces of water and oxygen, forming droplets, and then clouds, of sulfuric acid. As with clouds on the Earth, these lead to rain, but a rain of sulfuric acid rather than water. Actually this acid rain evaporates before reaching the extremely hot surface of the planet. Clearly the environmental conditions on Venus are horrendous: devastatingly high temperature, pressure, and winds, and acid rain—a far

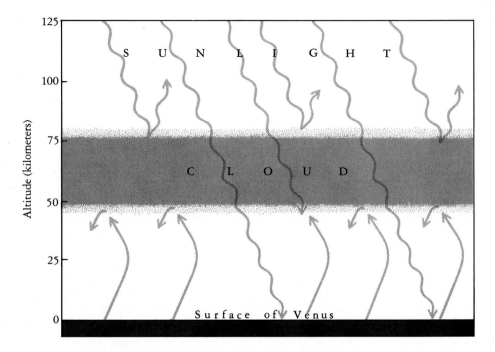

Mechanism of the greenhouse effect described in the text.

cry indeed from the beauty and charms implied by the image evoked by its name and from the early dreams of life on the planet. Yet, in spite of the thick atmosphere and clouds, a sufficient amount of diffused light reaches the surface of the planet to create visibility conditions similar to those on a cloudy, overcast day on Earth.

No one is surprised that at the present time Venus, with its surface temperature near 750 K, has no liquid water on it and only negligible amounts of water vapor in the atmosphere. Tom Donahue, of the University of Michigan, and his collaborators have concluded, however, that long ago Venus had an amount of water comparable to that now on the Earth, but that this water was decomposed by solar radiation and heat, the resulting free hydrogen escaping into space and the oxygen becoming absorbed by the rocks of the planet's surface. The reason behind this conclusion is the fact that, of the two kinds of hydrogen atoms that exist in nature—light and heavy—their proportion on the Earth is about 10,000 to 1, while on Venus it is only 100 to 1. This difference is attributed to the greenhouse effect and the resultant high temperature of Venus, which enabled the light hydrogen to escape the planet's gravity more easily than the heavy. If Venus did once have terrestrial-size oceans, then, under the circumstances just postulated, all the hydrogen contained in their water would have escaped into space in a period of about 280 million years—a relatively short time when compared to the 4.6 billion-year age of the planets. Inasmuch as there is no way for Venus to recapture this water it will remain forever inhospitable to life.

The atmosphere of Venus is so dense near the surface that a beam of light sent in some horizontal direction would keep being reflected from the upper, less-dense layers, much as radio waves travel around the curvature of Earth by bounc-

Venus' surface as it might appear as a result of the effect of the refraction of light by its dense atmosphere. There would appear to be several suns.

ing off a layer of charged (ionized) particles in the atmosphere. This mechanism is also similar to that involved in a mirage—or fata morgana—in the deserts on the Earth. As a result, a sufficiently strong beam of light sent in a horizontal direction on Venus would circle the planet, returning to the point of origin from the opposite direction. In principle, an observer could see the back of his head at a distance of some 40,000 kilometers. The poor transparency of Venus' atmosphere, however, makes it impossible to verify this fantastic conclusion.

To observe the surface of Venus through its opaque cloud cover, radar studies were done from the Earth and from orbiting spacecraft. These studies yielded maps showing the surface relief with considerable accuracy. The planet has mountains higher than Mount Everest, valleys, and canyons, the largest canyon being an enormous valley some 1600 kilometers long and nearly 5 kilometers deep—far bigger than our Grand Canyon. The radar maps show only a few large craters and no small ones, the latter because the resolution capability (that is, the

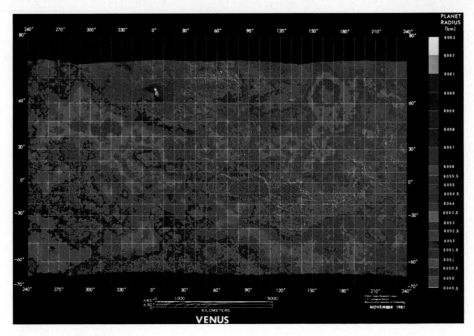

A radar relief map of Venus' surface. Colors indicate altitudes as shown in strip at right, which shows elevation as difference in radius of the planet.

smallest dimension distinguishable) of the radar systems was rather low or because the thick atmosphere protected the planet's surface from small meteorites, which would have burned upon entry. The hot erosive atmosphere and wind would also have tended to obliterate traces of older and of smaller craters.

We know very little indeed about the interior of Venus. The similarity to the Earth in diameter, density, and distance from the Sun imply a similar process of formation, so that we can expect a not too different interior chemistry and structure. In fact, some of Venus' interior may still be liquid, like Earth's, as is suggested by the presence of rocks of presumably recent volcanic origin. So far, there is no evidence of large-scale motion of parts of Venus' crust, such as the plate-tectonic movements of Earth's crust. The absence of a magnetic field is not surprising in view of Venus' very slow rotation—the slowest of all planets. We have noted that generation of a magnetic field requires that a planet have, among other things, a rather high rate of rotation.

Although present conditions on Venus appear to exclude the possibility of life there now or in the future, one cannot ignore the chance that, before the runaway greenhouse effect led to the very high surface temperature, there may have been a period when water and gases other than carbon dioxide were present and temperatures were comparable to those on the Earth. Unfortunately, the great difficulties in studying the surface features on Venus make it most unlikely that we will ever be able to find any fossils, the best evidence that there was once life on that planet.

CHAPTER 3 THE WATER PLANETS: EARTH AND MARS

CHAPTER 3

THE WATER PLANETS:
EARTH AND MARS

For years scientists have pondered why the Earth, whose orbit lies between and close to Venus and Mars, has an atmosphere and climate radically different from the other two—particularly in its abundant water, sine qua non of life as we know it. The deeper we look into this problem, the more fascinating it becomes. On the basis of recent spacecraft explorations, however, it appears that the Earth does not have a monopoly on water, and that Mars, the next planet further from the Sun, also has much of it, though only in frozen form.

EARTH

The Earth has four characteristics unusual for terrestrial planets: oceans, life, a strong magnetic field, and gradual motion of its continents (continental drift). The outstanding fact about the Earth is that there is life on it, while there are no signs evident of life on any other planet. We know also that life—the term will refer throughout to life as we know it—requires the presence of liquid water on the surface, of oxygen in the atmosphere, and of temperatures and pressures of a moderate and fairly narrow range. The question whether, under other conditions other forms of life could exist has, as yet, not been answered in a decisive manner though the answer is very likely negative. Thus we are led to examine the issue of why these fundamental requirements are satisfied on our planet and not on our neighbors Venus and Mars.

Preceding page: *Earth from the Viking 1 orbiter. The Arabian peninsula and a portion of eastern Africa are clearly visible in the upper portion of the picture.*

Were we strangers to the Earth speculating that, because of its location between Venus and Mars, its atmosphere and climate might also be "between" the limits represented by those planets, then we might expect the Earth's atmosphere to consist predominantly of carbon dioxide with a little water vapor and essentially no oxygen. We would also conclude that, on its surface, the average temperature was around 500 K (227° C) and the pressure 20–40 atmospheres. What a far cry from the actual atmosphere of about 80 percent nitrogen and 20 percent oxygen, an average surface pressure of one atmosphere, and a temperature near 290 K (18° C)! The last two are particularly important factors because they permit the existence of liquid water in the oceans. It is, therefore, clear that the evolution of the planet Earth was radically different from that of Venus and Mars and that the resulting unique characteristics made it possible for life to develop on it.

The available explanations of the radically different conditions between Earth and its neighbors are not complete, although the basic ideas appear to be understood. To begin with, when the planets were forming during the early evolution and cooling of the nebula, the first matter to condense was rocks, iron, and later, other minerals. It is also believed that when the resulting protoplanets were big enough and their gravitational fields strong enough, they attracted hydrogen and helium. Thus, these two gases constituted the so-called primary atmospheres of Mercury, Venus, Earth, and Mars. Later, when the Sun reached the stage where it began to emit powerful solar wind and radiation, it blew off an enormous amount of its hydrogen and helium, which, like a hurricane, swept through the Sun's neighborhood and carried off the primary atmospheres of those planets near it. As a result, the terrestrial planets were left with essentially no atmospheres. They did have, however, considerable amounts of internal heat produced by the process of accretion and by the decay of radioactive matter. The bigger the planet the more heat it had from both sources and the slower its loss into space. This heat was great enough to melt the interiors of these planets, and the heavier constituents, such as iron and other metals, drifted toward their deep interiors, leaving a less dense, rocky crust on the surfaces.

The gradual process of cooling and solidification of the initially all-liquid interiors of the planets was accompanied by the evolution of gases, which produced considerable internal pressures and eventually volcanism. Present-day volcanos emit lava and gases—mostly water vapor, about half as much carbon dioxide, and some nitrogen and sulfur. In the early history of the terrestrial planets, the volcanism was much more violent and widespread, and it has been suggested by American physical chemist and Nobelist Harold Urey that the gases emitted at that time contained more water and less carbon dioxide than now—as well as ammonia and methane. These gases were the basic constituents

Mt. St. Helens, in southwestern Washington, in eruption in 1980. Such volcanism was widespread in the early history of the planet.

of the new atmospheres of the terrestrial planets. Being small, Mercury had less internal heat than the other planets and, with its lower gravity, was unable to hold the emitted gases as an atmosphere.

At this stage the compositions of the new atmospheres of the terrestrial planets did not differ much from each other. With further evolution, however, profound differences began to be introduced. The main factor was the proximity of the Sun. Venus receives about twice and Mars about half the heat received by the Earth from the Sun. These relatively small differences became enormously exaggerated by greenhouse effects on the three planets. On Venus, a runaway greenhouse effect produced, as we have seen, temperatures at which most of the liquid water evaporated and was partly dissociated into hydrogen and oxygen. The first, being very light, escaped from the planet, and the latter became incorporated into the surface rocks. Because of the high temperatures, many surface rocks emitted gases that were initially entrapped in pores and cavities or, as carbon dioxide, were products of the rocks' decomposition. With liquid water gone, all that remained was a hot and dense carbon dioxide atmosphere, which, with small variations, exists today.

On Mars, the situation was the opposite of that on Venus. Its rather small size and, consequently, lower gravitational force did not lead to much internal heat, so that volcanism was weak and few gases came from degassing of the rocks. Whatever gases were present either froze on the polar caps and in the rocky soil

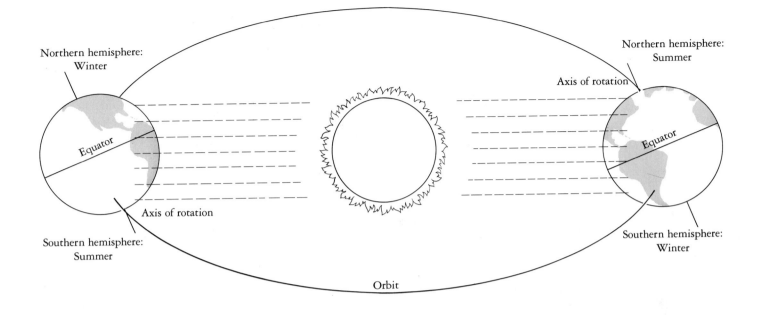

Northern hemisphere: Winter

Equator

Axis of rotation

Southern hemisphere: Summer

Northern hemisphere: Summer

Axis of rotation

Equator

Southern hemisphere: Winter

Orbit

The inclination of the Earth's axis and its role in the alternation of seasons in the Northern and Southern hemispheres. Note that the angle of incidence of the Sun's rays is more acute in winter.

of the planet's surface or escaped the small gravitational field of the planet. As a result, its atmosphere was thin and there was no significant greenhouse effect. Again, as with Venus, carbon dioxide remained as the main constituent of the atmosphere because the temperatures were low enough to freeze or liquefy most of the water. There was also very little oxygen.

On the Earth the situation was different and much more complicated. First, the water vapor of its secondary atmosphere gradually condensed and formed oceans. These oceans, in turn, absorbed huge amounts of carbon dioxide so that the atmosphere was no longer dominated by this gas. Solar radiation decomposed much of the ammonia and some of the remaining water vapor, producing oxygen and nitrogen. In the oceans, early life, especially green algae, like other plants, used carbon dioxide and released oxygen. Thus, the atmosphere of the Earth began to resemble what we have now, although in the beginning the oxygen content was so low that ozone did not form. As a result, for hundreds of million of years life could exist only in oceans, where the water provided protection from deadly solar ultraviolet radiation. Later, when the amount of oxygen increased, some of it turned into ozone, which formed a protective layer at the top of the atmosphere. From then on, plants and other forms of life developed, some of which could exist also on dry land. It is noteworthy that some of the carbon dioxide dissolved in the oceans was incorporated into shells of various species that eventually fell to the ocean bottoms, forming enormous deposits of limestone. In fact, it has been estimated that Venus and the Earth contain comparable amounts of carbon dioxide—on Venus in its atmosphere, on Earth, built into its limestones. Of central importance is the fact that the formation of our atmosphere, which is necessary for life, was a result of a close balance between several

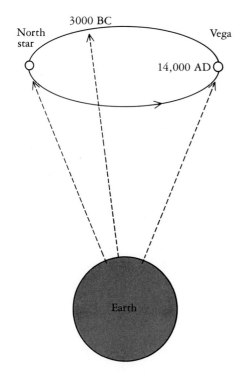

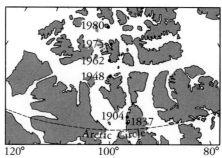

Locations of the Northern Magnetic Pole during the last century.

The slow but continuous change in orientation of the Earth's rotational axis, which resembles the motion of a spinning top.

independent processes. If one or another had been weaker or stronger, or had the timing been wrong, the Earth would have been uninhabitable. The existence of life on Earth makes it unique in the solar system.

The largest of the terrestrial planets, Earth revolves around the Sun in an orbit that deviates from a circle by just over three percent. Were it not for the 23°27' inclination to the plane of its orbit of Earth's axis of rotation there would be, at each latitude, only a seven percent variation of temperature during the year—either eternal spring, summer, fall, or winter. Like many things about the Earth, the axis of rotation is not fixed but slowly changes its direction among the stars. Thus the axis' northern end, which points roughly in the direction of the pole star, is drifting very slowly away from that direction. In about 26,000 years, it will once again point in its present direction. The reason for this so-called precession is the pull exerted on the Earth by the Sun and the Moon. The Earth is slightly oblate so that it has, like some of its inhabitants, a bulge at its equator. The precession motion is essentially like the wobble of a spinning top, the wobbles occurring 26,000 years apart. It has been calculated that if only the Sun were producing the precession of the Earth's axis the period would be nearly three times longer. Actually, the axis of rotation is not exactly fixed with respect to the surface, and so the true north and south poles wander a little in the region, as do the magnetic poles. Precise measurements of the orbits of the innumerable artificial satellites that circle the Earth have shown that it is slightly pear-shaped, the North Pole being some 20 meters higher than the average surface and the South Pole correspondingly lower. Also, the equator is not exactly circular, projecting 60 meters in the vicinity of the eastern Pacific and having an 80-meter depression in the Indian Ocean.

Another feature not as constant as we may think is the length of our day, which is getting progressively longer. The reason for this is as follows: The Moon, and to a lesser extent also the Sun, produces tides in the oceans. As the Earth rotates eastward, these tides tend to stay below the Moon, and thus move

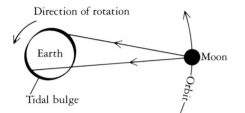

Direction of rotation

Earth

Tidal bulge

Moon

Orbit

Positions of moon-induced tides on the Earth. As a result of tidal friction the Earth tends to carry the tidal bulge with it, so that it does not occur along an Earth-Moon axis. This phenomenon is responsible for slowing of the Earth's rotation and for the gradually increasing distance between the Moon and the Earth.

westward. Ideally, a high tide produced somewhere in the open ocean should stay right below the Moon. Actually, because of friction and the eastward rotation of the Earth, the planet tends to drag the tide partly with it against the pull of the Moon. The result of these opposing forces is that the Earth's rotation is gradually slowed and, at present, the day is lengthening at a rate of 16 seconds per million years. Another result of these tidal forces is that the Moon is moving away from the Earth at about 4 centimeters per year, a rate easily measurable with radar. These changes are insignificant for us now, but 500 million years ago the day lasted 21 of our present hours, as confirmed by characteristics of the structure of the fossils of certain oceanic organisms from that period. Some time in the *very* distant future, both the terrestrial day and the lunar month will have the same length—of about two present months—and, at that time, the Moon will be about twice as far from the Earth as it is now. The Earth, because it is 81 times more massive than the Moon, exerts a tidal pull on it even more powerful than the Moon exerts on the Earth. However, since there are no oceans on the Moon, these consist only in a slight deformation of the Moon.

Although the commonly experienced short-term variations in the Earth's climate are rather unpredictable, long-term variations, such as periods of extensive glaciation, are better understood. These very slow changes are a result of alterations in the size and shape of the Earth's orbit around the Sun, in the inclination of its axis, and in the direction of this axis with respect to the stars. These changes are related to the fact that the Earth is under the gravitational influence not only of the Sun but also the other planets—in particular, Jupiter and Saturn. These two giants, which are jointly more than 400 times as massive as the Earth, are responsible for most of the deviation in the shape of the Earth's orbit, from almost perfectly circular to more elliptical orbit, and also for the variation in the inclination of the axis of rotation between 22 and 24½ degrees of angle. Each of these changes occurs over a different period: the shape of the orbit, every 105,000 years; inclination of axis, 41,000 years; and precession, 26,000 years.

All three factors, which affect the amount of solar heat reaching the Earth and its distribution over the surface, sometimes strengthen, sometimes counteract each other. Were the lengths of the three periods in some simple numerical ratio—say 1:2 or 2:3—the ice ages would occur with regularity. Since this is not the case, the severity and frequency of the cold periods and the occurrence of glaciation varies in a complicated though predictable manner. A careful study of the geological evidence for the last few million years confirms the theory originally proposed by Serbian geologist Milutin Milankovic in 1920. It appears that major ice ages occur 100,000 to 150,000 years apart, with many intervening minor ups and downs of the mean annual temperature. At present, after a period of high temperatures some 10,000 years ago, the Earth is heading

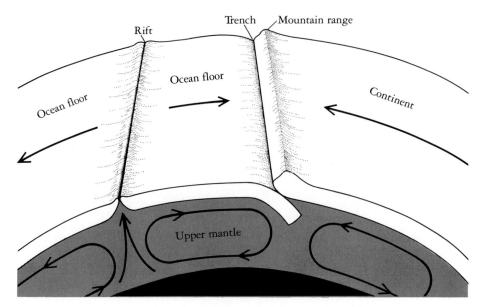

Convection in the easily deformable hot upper mantle accounts for motion of the plates that make up the Earth's crust. An upwelling of lava occurs at the rifts.

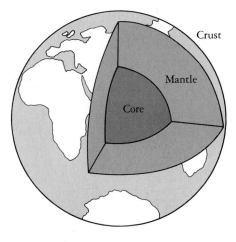

Schematic view of the Earth's interior.

for a very cold period in about 60,000 years, though there will be temporary smaller variations before that. There are, of course, short-term or local changes as can be easily seen from the fact that some glaciers are decreasing while others grow. It is suspected that the presence of mankind and the associated addition of carbon dioxide to the atmosphere have also had some effect on the climate. The importance of this effect is, however, still much debated. In general, the present Northern hemisphere has more severe cold periods than the Southern hemisphere; this is because the Northern hemisphere has more land which, when covered with snow, reflects sunlight back into space. The vast oceans of the South tend to absorb and retain the heat of the incident solar radiation. In the very distant past, the situation may have been different.

The interior of the Earth is fairly well known, primarily from studies of the reflection and refraction of natural or artificial shock waves, by various layers of different density. These phenomena are quite similar to that of the propagation of light through media of different density—as air and glass: some of it is reflected and some changes direction—that is, becomes refracted. It is certain that, from the crust to about half way to the Earth's center, there is a rocky, mostly solid region called the mantle. Below that there is a core, which consists of liquid outer and solid inner parts. Both are made of iron and nickel and their compounds, which, because of pressure about a million times atmospheric pressure on the Earth's surface, are at least twice as dense as the iron and nickel that we know in daily experience. The reason some deep layers are solid and others liquid is that melting temperatures usually increase with pressure, and both pressure and temperature are higher at greater depths. The interior of the Earth is hot because of the energy released by radioactive elements; the temperature at the center is estimated to be between 3000 K and 10,000 K, which, interest-

A "black smoker," or hydrothermal vent. The black sediment-laden plume is a product of the upwelling of mantle material that occurs at underwater rifts.

ingly enough, is not far from the temperature of 6000 K of the Sun's visible surface although there is no direct connection between them.

One consequence of the heat is the relatively easy deformability of the inner, hotter layers of the mantle, which are so soft that there are slow convective motions in which the hotter parts rise and the cooler parts sink. The crust is broken into many plates, which float on this partly convective mantle, moving slowly together or apart or past each other. Those parts of the plates that rise above the oceans are the continents. The idea that continents move, first proposed in 1910 by German meteorologist Alfred L. Wegener, met with enormous skepticism. Now we know many facts that tend to confirm these motions—that, for instance, the two Americas are moving away from Europe and Africa at an average rate of two or three centimeters per year. Deep in the ocean, through the rift where these plates separate, hot liquefied rock wells up and solidifies, to form the mid-Atlantic ridge, a submarine mountain range. Tracing back this motion—that is, to the point in time where there was no Atlantic Ocean, shows that the coastlines of these continents fit together very well. Similarly, it can be shown that all continents can be fitted together in this fashion, so that it is quite certain that 150 to 200 million years ago there was only one huge continent (designated Pangaea) surrounded by one enormous ocean. Later this primitive continent gradually broke into individual plates and continents.

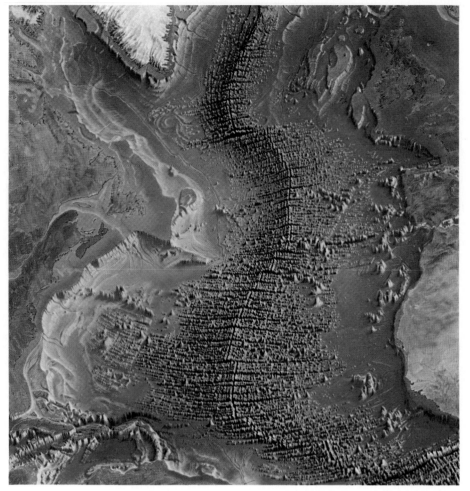

The Mid-Atlantic ridge, boundary between eastward- and westward-moving plates. The rate of movement is 2–3 centimeters per year.

Pangaea, the single land mass of about 200 million years ago, which later broke up to form the individual continents.

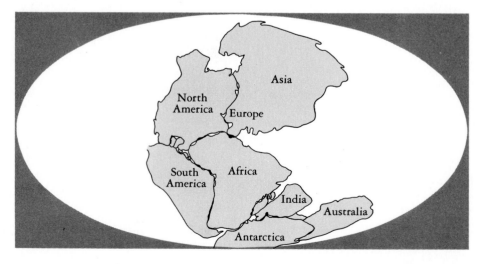

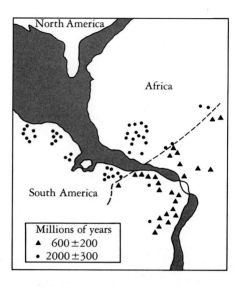

Correspondence in age of rocks from Africa and South America is evidence of the origin of the continents in a single land mass.

Another strong indication of the validity of the idea of continental drift is the patterns of magnetic orientation found in the rocks on the two sides of a rift, which show their orientation with respect to the Earth's magnetic field, thus providing an indelible record of the origin and direction of motion of the ocean floor and, hence, of the continental plates. Also supportive of the theory is the correspondence mentioned above of coastlines and, in addition, of various geological features and formations—such as those along the eastern shore of Brazil and the western shore of Africa, which fit together like the parts of a jigsaw puzzle. There are, also, striking similarities among animals. For instance, as pointed out by Wegener, marsupials such as the kangaroo, which have a pouch in which they carry and feed their young, are found mainly in Australia and in South America, implying some earlier connection between the two by way of Antarctica, in which marsupial fossils were recently discovered. Similarly, flightless birds appear to exist only in Africa (ostrich); Antarctica (penguin); Australia, New Guinea, and East India (cassowary); South America (rhea); and New Zealand (kiwi)—implying again a land link among these areas. Other such evidence implies a similar link between Scotland, Iceland, the Faeroe Islands, Greenland, and the Canadian Arctic.

Even casual observation of the sides of valleys and canyons cut by rivers or glaciers or of highway and railroad cuts suggests that mountains were formed by folding and bending of the Earth's crust. Were the Earth's interior solid, such

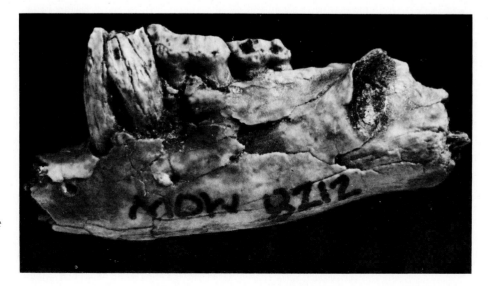

Fossilized jaw fragment, found in the Antarctic, of a marsupial that lived at least 40 million years ago may be evidence of an earlier land connection between Australia and South America.

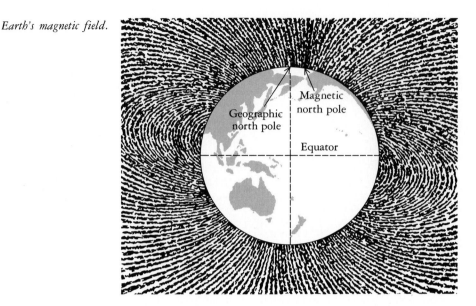

Earth's magnetic field.

Geographic
north pole

Magnetic
north pole

Equator

bending would be nearly impossible. Since, however, the crust is relatively thin and the plates actually float on the hot and liquid interior, they have been and continue to be deformed. Where the plates collide one usually slips under the other and eventually melts in the hotter, deeper layers of the mantle. Where the plates separate, liquefied rock fills the gaps. In this way, the plates and continents are being slowly and constantly renewed. It is not surprising that a very close correlation has been found between the boundaries of the plates and the areas of occurrence of earthquakes—as on the western coasts of North and South America or in Japan. A separation of plates is well illustrated by the Red Sea, which was formed as a rift between the African and an Asian plate. Another famous effect of plate tectonics is the northward push of the India plate against the Asian plate. This motion presumably formed the Himalayas and is related to the severe earthquakes in China and in the vicinity of the enormously deep rift in Siberia that is now filled with water and called Lake Baikal.

Very recently, Clark Blake and Porter Irwin of the U.S. Geological Survey have shown that, some 200 million years ago, the western coast of North America lay east of the High Sierras, along the present border of Nevada down to Death Valley and the Mojave Desert. Over the next millions of years, various chunks of continents and islands were carried, by plate tectonics, northeastward across the equator until they crashed into this early western coast. There they built new mountain ranges and created valleys, the last such addition having been made some 15 million years ago. It is expected that these tectonic move-

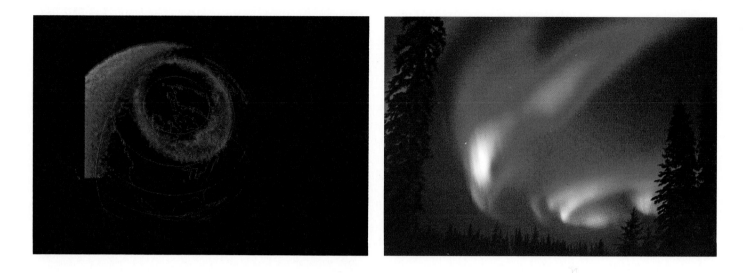

At left, view of an aurora taken by a spacecraft at an altitude of about 20,000 kilometers. This image was made at ultraviolet wavelengths. At right, auroral bands in the Alaska sky.

ments, evidenced by the relatively frequent earthquakes along the famous San Andreas fault, will continue for the next 50 million years.

A promising tool for the observation of plate motion and, thus, for the possible prediction of earthquakes, is the use of radar reflections from the Moon. The Moon provides a steady, reliable and accessible reference point, and radar measurements of its distance can be correct to within a few centimeters—comparable to the extent of the annual motion of plates or to other large-scale tell-tale deformations of the Earth's surface.

A familiar feature of the Earth is its magnetic field. Magnetic fields in everyday experience are generated by electrical currents flowing through a loop or coil of wire. In Earth's field the current flow is in its metallic liquid core. These currents are sustained by convective motions between the hotter and cooler parts of the liquid and by other motions produced by the rotation of the Earth. A quantitative evaluation of the energy of these processes, however, seems to indicate that they are marginal, so that other mechanisms have also been proposed. As a result of the role of rotation in generating the field, its axis is almost parallel to that of the magnetic field, so that the magnetic North Pole is not far from the geographic North Pole. Since the motion of a liquid, here the liquid core of the Earth, is rarely very regular, it is not surprising that the magnetic pole is not fixed but moves slightly in an irregular manner. There is, in fact, evidence that 100,000 years ago the polarity of the Earth's field was the reverse of what it is today and that there have been many such reversals over the years.

An important consequence of the presence of a magnetic field around the Earth is that charged particles, such as electrons and protons, are trapped by the

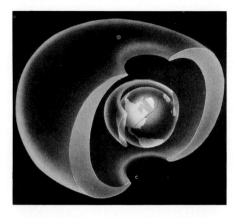

The Van Allen belts, composed of charged particles that surround the Earth at altitudes that vary between 2,000 and 16,000 kilometers.

field in wide doughnut-shaped belts surrounding our planet at an altitude of 2000 to 16,000 kilometers. The particles moving in these belts, called Van Allen belts after their discoverer James Van Allen of the University of Iowa, produce electromagnetic radiation that can be easily observed. The presence of such radiation is a very convenient indicator, in the study of distant planets, of whether a particular planet has a highly conductive and liquid interior—and, in consequence, a magnetic field.

The temperature of the Earth's atmosphere shows striking variation with altitude. It is about 200 K ($-75°$ C) at an altitude of 10 kilometers. From there to 50 kilometers it increases more or less uniformly to 273 K ($0°$ C), after which it begins to drop, falling to 173 K ($-100°$ C) at 100 kilometers. From that elevation upward, there is a steady increase up to a few thousand degrees. These odd ups and downs can be accounted for quite well by various chemical reactions and solar heating. Only half of the incoming solar heat reaches the Earth's surface, 15 percent being absorbed by the atmosphere and 35 percent reflected back into space, mainly by clouds.

The Earth's atmosphere places a definite limit on the usefulness of the best modern optical and infra-red telescopes. Some of these limitations can be minimized by observing from high mountain tops such as Mauna Kea (4215 meters) in Hawaii or Pic du Midi in France (2861 meters), but even at such high altitudes the air contains water vapor, which absorbs radiation coming to us from the stars. There is also the unavoidable problem that air is never at rest, with small-scale changes in density and temperature occurring continuously. These changes lead to distortions in the image not unlike those we often see in the heat-shimmer above a distant highway on a very hot day. These problems will be at least partly solved by the Space Telescope, which will be placed in orbit around the Earth by the space shuttle. At these altitudes, a telescope of just under 2 meters in diameter will be many, many times more effective than the much bigger ones on the ground.

There is recent fascinating evidence that radiation generated in supernovae explosions that occurred many, many light years away left permanent traces on the Earth. Although this radiation, in the form of high-energy X-rays, falls on all of the Earth, only the slowly accumulating ice layers in the polar regions remain sufficiently unperturbed for centuries on end to preserve a record of it. That record consists of sharply delineated thin layers of high concentration of a kind of nitrous oxide that is formed by very high-energy radiation. Knowing the rate at which the ice was deposited, we can deduce that these layers were formed around the years 1600, 1575, and 1180, dates that correspond closely with those when unusually bright supernovae were observed—1604, 1572, and 1181. The polar ice samples were not taken from levels deep enough to determine whether

Path of the Earth

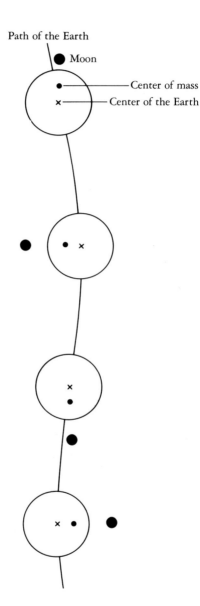

Moon

Center of mass
Center of the Earth

The mass of the Moon is a non-negligible fraction of the mass of the Earth. Hence, although the center of mass of the sum of the two moves along the orbit around the Sun, each of the two bodies follows a slightly wavy path.

the famous supernovae of 1054 and 1006 also left traces, but expectations are that they did.

THE MOON

It has been said that the main stimulus in the very early study of the sky and of the stars was the presence and the obvious motion and variability of the Moon, including eclipses. The ratio of the Moon's mass to the mass of the Earth is greater than that of any other satellite and its planet with the possible exception of the Charon-Pluto pair. To begin with, we have to admit a most embarassing uncertainty about the Moon's origin. Stephen Brush of the University of Maryland, designates three kinds of theories: the "sister," "daughter," and "wife" theories, which, respectively, assume that the Moon was accreted from a portion of the solar nebula near where the Earth was accreted, that it was formed from a part of the Earth that broke away when the Earth was still partly liquid, or that it was formed elsewhere and was captured by the Earth. All three proposals have their proponents and opponents and all three have strong and weak points. The first seems to be a natural extension of the process of the formation of the Earth, but it appears virtually incapable of explaining why the dust of the solar nebula accreted and formed the huge Moon near the Earth rather than falling directly onto it, and why the chemistry of the Moon is significantly different from that of the Earth. The second theory has shortcomings with respect to providing a reason for and the mechanism of the breakaway. The suggestion that the Pacific is the scar left by the breakaway appears to be untenable in view of the effects on the Earth's crust of the forces of plate tectonics, which would have obliterated all traces of this cataclysmic event. The attractiveness of this proposal lies in its seemingly natural explanation for a certain similarity between the Moon and the Earth and for their proximity to each other. From some points of view, the third alternative is the most plausible, but here also there are serious deficiencies in explanation of the process of capture. As a body approaches the Earth, or any planet, from far away, its velocity and kinetic energy increase. Unless the body is on a direct collision course it will pass the Earth, just as a comet passes the Sun, and drift away. The only way the body can be drawn with in a closed orbit around another one is for the excess velocity and energy to be lost in some manner. This can be brought about, for instance, by interaction with another satellite or by drag produced on the body by a ring of dust or dense gas. So far as we know, however, the Earth had no earlier satellite and any drag exerted by whatever was left of the Earth's nebula would have been much too weak to affect so massive a body as the Moon. The mass of the Moon is an important consideration because we know that small bodies can be forced into an orbit by such weak

The nearside of the Moon. The farside has many craters, but relatively few maria (the flat dark areas).

At left, an Apollo astronaut on the Moon near a boulder that had rolled down from a nearby slope; at right is the moon vehicle, or rover, used by the astronauts. Note the footprints in the fine powdery "soil" of the Moon.

Giordano Bruno, a bright-rayed crater on the farside of the Moon. The brightness of the rays indicate that they are of newer material than their surroundings.

interactions. Recently it has been suggested that the Moon may have had a grazing collision with the Earth, but no evidence of this has been found. A serious problem for all theories of the origin of the Moon is posed by its composition. Although the Moon's average density is comparable to that of the outer parts of the Earth, there are substantial differences in the concentration of certain elements such as titanium, potassium, and carbon—facts established when the Apollo missions made the Moon the first extraterrestrial body on which man was able to walk, and, thereby, bring back to the Earth 1800 kilograms of rock samples.

The nearside of the Moon is a familiar sight, with its dark flat areas, called maria—"seas," which they were once thought to be—and huge impact craters. The farside is mountainous and has only a few small maria. Measurements of gravity by a Moon orbiter indicate that the lunar crust on the nearside is relatively thin—more than 50 percent thinner than the crust on the farside. This phenomenon could have been brought about by the impact of huge bodies and, much later may have allowed molten lava to emerge and form the maria. Study of the radioactivity of rocks brought back by Apollo and the Soviet Luna missions, has made it possible to state with considerable certainty that most of the cratering occurred 4 billion years ago, the basaltic rocks of the maria being somewhat younger. The oldest rock so far found is 4.4 billion years old.

An important result of a study of the size and number of craters on the Moon is that there is a close relationship between the two—one that holds fairly well for other bodies in the solar system (as long as they have no atmosphere to prevent penetration to the surface of small meteorites and to cause erosion) all the way from Mercury to the satellites of Saturn. From estimation of increases of temperature with depth below the Moon's surface, from comparison to conditions on the Earth, and from study of the rate of the Moon's very small oscillations around its Earth-facing orientation, it has been concluded that, if part of the Moon is liquid, it is probably limited to the central core, which is less than 1000 kilometers in diameter. Of course, by now the Moon may be completely solid, a possibility supported by the nearly total absence of an overall magnetic field, although there are areas on the surface that show some local magnetism. Seismometers left on the lunar surface by Apollo astronauts show that all the moonquakes recorded can be explained by impacts of larger meteorites in their neighborhoods or by a slight cracking of the lunar surface when the tides raised in the crust by the Earth become particularly strong.

The lunar surface was, of course, bombarded for billions of years by the protons and electrons of the solar wind. The result of this bombardment and of other factors is that freshly exposed surface material darkens and that, as predicted by the author, the very fine particles composing the surface dust adhere to

The crater Tycho, about 100 kilometers in diameter. Note the central mountain peak and smaller craters on its floor.

each other and form a thin crust. The older lunar surface has a reflectivity comparable to that of charcoal, which is why the more recent craters are so much brighter than the old surface. The huge dark Mare Imbrium, which is some 800 kilometers across and is easily visible with the unaided eye in the northern part of the Moon, is about 4 billion years old. The bright Copernicus crater south of Mare Imbrium is 1.1 billion years old, while the big Tycho crater, fairly far out to the South, is only 100 million years old. The small South Ray Crater was made about 2 million years ago—that is, at a time when man already walked the Earth.

The center of gravity of the Moon is slightly displaced from the geometrical center of the body toward the lunar nearside. As a result, the Moon rotates in such a manner that it always turns the same face toward the Earth. Actually, the Moon also performs a slow rocking motion, not unlike a kite on a string, with the result that from the Earth one can see 59 percent of its surface.

MARS

The most striking aspect of Mars, besides its fantastic mountains and canyons, is the seeming evidence that in the past it had liquid water and rivers and that it could be said to have had, at that time, a friendly climate, hospitable to life.

Mars was noted very early by man because of its distinctly red appearance, in acknowledgment of which it was named for the god of war. Later on it was found

The looping path of Mars' retrograde motion.

An oblique view of a portion of Mars, with the horizon about 19,000 kilometers away. A thin haze accounts for the brightness of the horizon, 25 to 40 kilometers above which are other layers of haze that may be composed of crystals of carbon dioxide.

that it had a peculiar motion with respect to the stars: while other planets that are farther from the Sun than the Earth move from east to west in our sky, Mars seems to stop occasionally, reverse its motion, stop again, and then continue on its regular westward way. The explanation of this odd behavior—called the retrograde loop—is that the Earth moves along its orbit about twice as fast as Mars, which is one and a half times as far from the Sun. The situation is comparable to riding in a train that overtakes a slower-moving car on a parallel road, which then appears to the rider to be moving backward with respect to the distant horizon. Actually, all outer planets exhibit similar periodic anomalies of motion, but that of Mars is particularly large and easily observable.

For many reasons, Mars was fascinating to astronomers: its day is about as long as ours, the tilt of its axis is about the same as the Earth's, its trip around the Sun is only about twice ours, and it has dark regions and white polar caps, all of which show seasonal variations in size. It was early and correctly concluded that Mars has an atmosphere. Altogether the planet appeared to have a great similarity to the Earth and so its exploration by spacecraft had a high priority, both in American and Soviet space programs. It was thus a big disappointment

The "canals" of Mars, as seen in 1896–7 by Percival Lowell, founder of the Lowell Observatory.

The north polar ice cap of Mars, showing patches of water ice in the upper part, with layering (light and dark banding) and surrounding dune fields. The ridging at lower left is composed of sand.

when, shortly after Mariner 9 spacecraft left on its six-month flight to Mars, the surface of the planet became covered with a huge opaque cloud of dust so that no features of the planet could be seen. This cloud turned out to be a blessing in disguise, however, because its slow abatement and settling down was observed in detail and revealed details of the surface relief never dreamt of before. These high-altitude observations and the results obtained by other orbiters, and especially landers, produced a body of knowledge about Mars surpassed only by our knowledge of the Earth.

For decades the seasonal changes visible on Mars were interpreted as a sign of seasonal growth and decline of vegetation. This natural tendency to interpret these observations in terms of conditions on the Earth culminated in the assertion by respectable observers, such as Percival Lowell, the founder of the Lowell Observatory in Arizona, that they saw on Mars a network, hundreds of kilometers long, of perfectly straight canals. Once these canals were accepted, it was easy to invent Martians with technological skills adequate to travel to the Earth in UFOs. All these imaginative conclusions collapsed when better telescopes were built and especially when various spacecraft revealed the true nature of the planet's surface.

Although there is water on Mars in the form of permafrost and some water vapor and clouds, we know now that the polar caps are made mostly of solid carbon dioxide, familiar to us as "dry ice." The changes in the darkness and size of the dark regions have proven to be the result of enormous seasonal winds that transport surface dust, periodically covering the dark areas with bright dust and then stripping them bare. Apart from minor components, the atmosphere of Mars is almost pure carbon dioxide at a pressure about 0.5 percent of that on the Earth. The mean surface temperature is some 45 K lower than that on the Earth; on the Martian equator it varies from an occasional 300 K (27° C) to 175 K (−100° C); polar temperature may drop to 135 K (−140° C). The mysterious canals turned out to be an optical illusion—the product of a human perceptual tencency to see local regularity in a more or less random array of shapes and spots. We know now that these spots and shapes are craters, curved channels, and other geological forms.

A mystery comparable to that of the straight canals concerns the origin of long, irregularly curved channels or depressions, most of them near cratered areas. These channels look very much like dry riverbeds, such as the Southwestern arroyos secos or African wadis, even to the inclusion of teardrop-shaped islands—that is, islands apparently shaped by flowing material. Efforts to explain these channels as having been formed by wind erosion or by lava flow appear to be untenable, and so it is tempting to think of them as dry river beds. The mystery arises when one considers the amount of water necessary to form

Drifts of abundant bright dust on the surface of Mars. Shifting of this dust by enormous seasonal winds accounts for the periodic changes in the bright and dark regions of the planet. (View continues on facing page.)

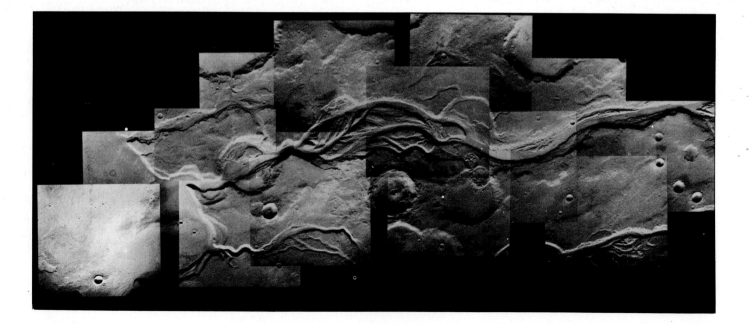

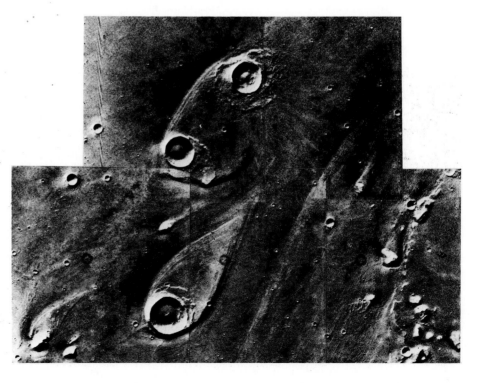

On facing page, riverbed-like channels, and at right, teardrop-shaped islands typical of those found in rivers on Earth. Both features suggest water erosion.

A detail of the polar ice cap of Mars shown earlier, with layering more clearly visible. The layers may indicate varying climate.

channels that are 15–30 kilometers wide, hundreds of kilometers long, and up to one kilometer deep. All these dimensions are larger than those of the Amazon, and the required flow of water would be 100 million to one billion cubic meters per second, or 10,000 times the rate of flow of the Amazon. The question raised by this puzzle is important because, were there ever water in that amount on a planet in so many respects similar to Earth, then there may have been life on it.

The present atmospheric pressure and temperature on Mars are such that water cannot be present as a liquid but is frozen in the form of underground permafrost—analogous to those in Alaska and the Canadian Northwest territories. If water is indeed present in this form, then one has to explain why millions of years ago the planet's climate was so much warmer—warm enough to permit the occurrence of liquid water.

We know from long-range climatic changes on the Earth that Jupiter pro-

A portion of the crater of Mars' volcanic mountain Olympus Mons, which may still be active. The diameter at its base is 500 kilometers.

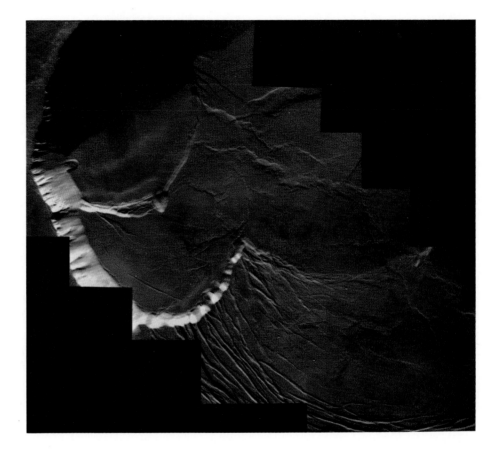

duces gradual changes in the Earth's orbit and in the inclination and orientation of its axis. For Mars, which is nearer to Jupiter than the Earth, one could expect these effects to be much more pronounced and, indeed, the calculated periodic changes in the inclination of Mars's axis are nearly ten times as large, and the changes in the shape of the orbit itself more than twice as large, as for the Earth. Furthermore, the planet's equatorial bulge is much larger than the Earth's, and so precession is stronger.

Was the Martian climate, then, much warmer? According to theory, periods of a climate some 30 K warmer than now have occurred on Mars and will occur every few million years, permitting some of the permafrost to melt. We do not have a solid geologic confirmation of this conclusion, although there are layered structures of dust and water ice in the polar regions, which suggest that major climatic changes did indeed take place. Another, less likely, possibility is that

Evidence of the action of the wind on Mars. Letters on lower picture indicate features concealed by dust in the upper (earlier) photograph.

there was on Mars enough volcanic activity to melt the permafrost temporarily. This melting would be, however, a local effect. If water was indeed present on Mars in the past there may have been not only huge rivers, which formed the many presently visible channels, but also a much denser atmosphere. A recent study of the nitrogen and noble-gas content of the Martian atmosphere indicates that Mars once had an atmosphere much denser than today's. Under these conditions, there may also have been life in some primitive form. It is true, however, that no traces of present or past life have been found; our exploration of Martian geology and of the planet's surface, however, did not include micromineralogical studies. It will take another mission to Mars if we are to have better answers to this tantalizing question, and Mars undoubtedly will be the first, if not the only, foreign planet on which man will walk.

Mars has many other fascinating and unusual features. Probably one of the most spectacular is the volcanic mountain named Olympus Mons. Three times higher than Mt. Everest and nearly 700 kilometers wide at its base, it is the

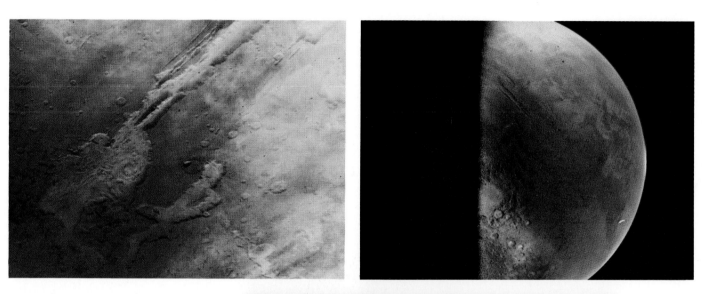

Above left, about one-third of the 5000-kilometer-long Valles Marineris. At top right, the huge canyon clearly visible as an enormous groove running diagonally across the face of the planet. At right is a close-up of a landslide that occurred in the 3-kilometer-high wall of the canyon.

largest known mountain in the solar system. Its shape is that of a regular volcanic cone, with a huge crater some 80 kilometers across. The volcano was probably active for a billion years and may be so even now. Calculations show that the weight of this and of other volcanic mountains in the nearby Tharsis area is so huge that, in order to support it, the underlying crust must be much thicker than that of the Earth. The fact that the diameter of Mars is about half that of the Earth suggests that whatever is left of its heat of accretion and radioactive heat is much smaller, and the rate of cooling higher, than those of the Earth; therefore, its interior may be entirely solid. On Mars, there are, of course, many impact features; sixteen are more than 250 kilometers across. On the whole, there are fewer small craters than on the Moon, probably because of the protective atmosphere, which may once have been much denser, and also because the highly mobile dust covers and erodes older surface irregularities.

Another spectacular feature on Mars is Valles Marineris, a canyon—or rather a system of canyons—up to 4 kilometers deep and 4000 kilometers long, a distance comparable to that between Washington, D.C., and San Francisco, or across the whole of Australia. While its similarity to the Grand Canyon suggests erosion by water, the proximity of its west end to the enormous Tharsis bulge implies that the volcanic mountains may have had a role in its formation—through the stresses and tension they produced in the crust. In some places, the walls of Valles Marineris have collapsed, the rocks sliding to its bottom; in other parts they are nearly perpendicular and provide an insight into various deeper layers of the Martian crust. The southern hemisphere of Mars has many mountains and craters, while the northern hemisphere is smooth, with very few craters; the origin of this difference is not understood.

Mars is the only planet whose surface was investigated in some detail by landers, which were delivered to it by two Viking spacecraft, each combining an orbiter with a lander. On first consideration, the parachuting of a lander onto a planet that has an atmosphere, might seem an easy matter. Actually, the choice of suitable and safe landing sites, in which the author was partially involved, was quite difficult. The best photographs of the terrain could not reveal the presence of boulders or even of irregularities the size of a small house. Clearly, a bad landing of a four- or three-legged instrument, itself the size of a very small house, was risky. On the basis of various physical and geological data sites were chosen that were most likely to be smooth. Luckily, both landers touched down safely, although the locations proved to be both similar and desolate.

The landers provided hundreds of pictures—not to mention climatic and meteorological data—for more than two years. The matter surrounding the landing sites ranges from rocks about one meter across to small pebbles and fine dust. It was possible to follow changes produced by wind and to see the forma-

At right, a view of the Martian surface taken by Viking Lander 2; below, a closer view of the surface, with the lander's soil sample collector arm visible at upper left and, at center, a rock moved by the arm and the depression left by the rock.

Traces of white frost seen on Mars' surface, generally in areas in which shadows occur. Parts of the Viking Lander 2 are visible in the foreground.

tion of Martian frost during the local winter in the northern hemisphere, where, because of an eccentricity of the planet's orbit more than five times that of the Earth's, winters are shorter and milder than in the southern hemisphere. The famous red color of Mars is caused by iron-rich rocks and iron oxides, as had been surmised in pre-Viking days. Many of the rocks appear to be pitted and severely eroded. In spite of being some 2500 kilometers apart, both landing sites appear to have comparable surroundings.

During the Viking missions samples of Martian soil were analyzed by automated devices for the occurrence of chemical reactions and of organic molecules considered essential for life as we know it on the Earth. Within the limits of the experimental technique no such evidence was found.

The two satellites of Mars, god of war, are wryly named Phobos (fear) and Deimos (panic). They are very small and were not discovered until some 100 years ago. Actually in 1726 Jonathan Swift, in *Gulliver's Travels,* describes Gulliver's visit to the imaginary land of Laputa, where Laputian astronomers observe two satellites of Mars, one circling the planet in 10 hours, the other in 21½ hours. In 1877, 150 years later, Asoph Hall, working at the U.S. Naval Observatory, discovered that Mars indeed has two satellites, which were totally unobservable with the simple telescopes in existence during Swift's life. What is more, one of the satellites orbits the planet in 7½ hours, the other in just over 30 hours—in remarkable agreement with Swift's description. Being familiar with the laws that govern the motion of planets and satellites, derived around 1600 by Kepler, Swift was able to calculate the appropriate distances from Mars of his two imaginary satellites. Swift's prediction was very likely based on Kepler's speculation that, since Venus has no satellites, Earth has one, and Jupiter four major satellites, then Mars should have two and the "missing" planet between Mars and Jupiter would have had three.

Like all small objects in the solar system, Phobos and Deimos are very irregular in shape. Phobos, the larger of the two, is so close to Mars that it circles the planet in only 7 hours and 40 minutes, more than three times faster than the planet rotates. Deimos, on the other hand, more than three times farther from Mars, revolves more slowly than the planet's rotation. As a result, during one Martian 24½-hour day, Deimos appears to move in the Martian sky slowly westward while Phobos dashes three times around the planet in an eastern direction. To someone standing on the surface of Mars—as we may yet—the apparent size of the rapid Phobos will be about half of the apparent size of our Moon, while Deimos will be 20 times smaller. As with our Moon, each satellite rotates at the same rate that it circles the planet, and so always turns the same face towards Mars.

Both Viking orbiter spacecraft provided more than 125 images of Phobos,

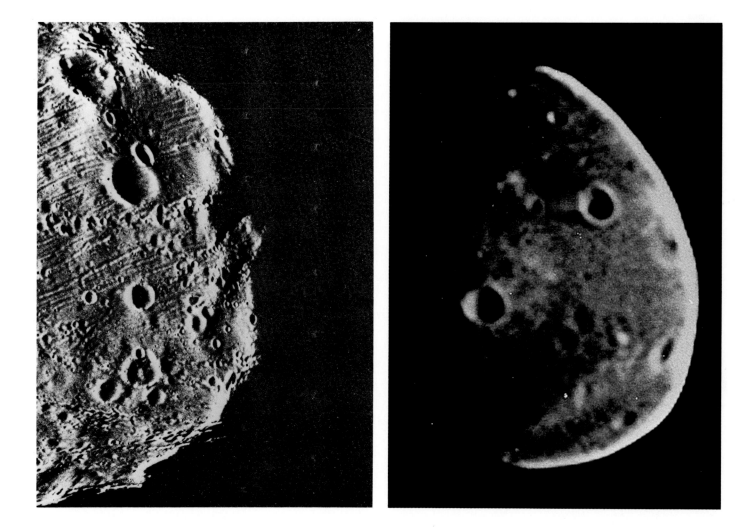

Mars' two satellites, Phobos (above) and Deimos, which are probably captured asteroids. The distance from top to bottom of the Phobos photograph is about 18 kilometers, and objects as small as 40 meters across are visible.

some at a resolution better than 30 meters. This satellite appears to be covered with sharp, crisp craters as small as 20 meters, some forming chains. It also has one huge, 10-kilometer crater, Stickney, which is associated with a vast network of parallel striations or grooves. The origin of these striations, which do not seem to occur on any other body in the solar system, is unknown. One proposal is that they may be the result of an enormous shock, or impact, perhaps connected with the formation of Stickney. Deimos, in contrast to Phobos, has so few small craters that it looks rather smooth, although it has a few large ones. Densities of both satellites are much lower than that of Mars, and it has been suggested that they may be captured asteroids. An argument in favor of this supposition is that the solar light reflected from these two satellites resembles the light reflected by some of the asteroids, implying that they may have a similar, if not identical, chemical composition.

Early observations of Phobos indicated that it is gradually slowing its daily trip around the planet, and it was conjectured that this must be due to the drag exerted by the atmosphere—admittedly very thin—at an altitude of some 6000 kilometers above the planet's surface. A simple calculation showed, however, that if this were the case, then Phobos must have a density at least 1000 times lower than that of water, and so according to the Soviet astronomer Shklovskii, must be hollow. In 1966, Shklovskii made the tongue-in-cheek proposal that Phobos, and perhaps Deimos were artificial satellites made by some superior civilization existing on Mars. Fortunately, it turns out that the progressive approach of Phobos to the surface of Mars is caused not by atmospheric drag but by a tidal effect and that Phobos will exist at least another 100 million years. Phobos and Deimos are, undoubtedly more like golf balls, than ping-pong balls.

CHAPTER 4   THE GIANTS: JUPITER AND SATURN

CHAPTER 4 THE GIANTS: JUPITER AND SATURN

If, as some suspect, there are in the universe more advanced civilizations than ours, and if these civilizations are able to study our solar system better than we can theirs, then it is likely that they notice, at first, only two of our planets: Jupiter and Saturn. These two spectacular giants constitute more than nine-tenths of the mass of all the planets together; from this quantitative point of view, the other planets do not amount to much. Jupiter and Saturn have been observed in great detail within the last decade by the extremely successful Pioneer and Voyager spacecraft and prove themselves as intriguing to us as they might seem to distant observers.

JUPITER

One of the seven wonders of the ancient world was the statue of Jupiter by Phidias. In the present world, the planet Jupiter is an equal wonder of the sky, sometimes the brightest "star" we see. Calculations of the thermal history of the early planets show that when Jupiter was formed from the solar nebula, it was even more spectacular—much hotter, much brighter and about ten times as big as it is now. The temperature in the center of the planet was then near 50,000° K and now, a few billion years later, it is still about 30,000° K. Jupiter's brightness also dropped during that time from being initially about one percent of that of the Sun to nearly one-billionth, a decline by a factor of 10

Preceding page: Saturn and its rings, the latter casting their shadow on the planet.

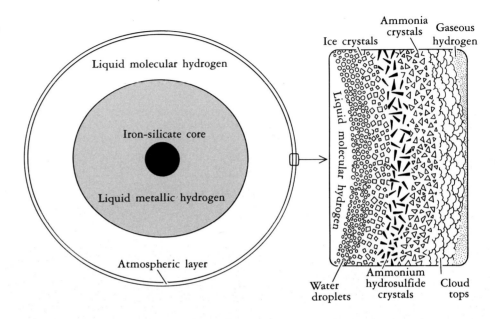

Structure of the interior of Jupiter as deduced from observations of the planet and its satellites; at right, the atmosphere of Jupiter.

million. Although the planet is gradually cooling by radiating heat into space, it is still very hot, giving off nearly twice the heat it receives from the Sun. By contrast, the small terrestrial planets have only a tiny supply of internal heat, produced by radioactive elements.

Although Jupiter's mass is one-thousandth that of the Sun, it is more than 300 times the mass of the Earth. In fact, Jupiter's mass is nearly three-quarters the mass of *all* the other planets put together. The suggestion has often been made that initially Jupiter and the Sun were unequal partners in a binary star system. Actually, if Jupiter's present mass were some 70 times bigger, or its diameter some four times larger, its central temperature would be high enough to start a nuclear reaction, and it would be a true star. It is perhaps fortunate for us that the Sun and Jupiter did not "make the grade" as a double star, for then the planets would have formed in quite different orbits, with consequently different temperatures and chemistry. In fact, it is very likely that Earth-like planets would not have formed and that there would have been no life in the solar system at all. Actually, we do not know whether any of the thousands of double stars we see in the sky have planets around them.

An enormous and magnificent planet, Jupiter is surrounded by powerful fields and radiations, some of which even reach the Earth; it is indeed king of our planetary family. Surprisingly enough, in spite of Jupiter's enormous size and its great distance from us, about five times our distance from the Sun, we probably know more about its interior than about the interiors of the terrestrial planets—

Banding of Jupiter's ammonia clouds, with the Great Red Spot visible at left. Two of Jupiter's satellites can be seen—Europa at right and Io at left, the latter partially obstructing the Great Red Spot.

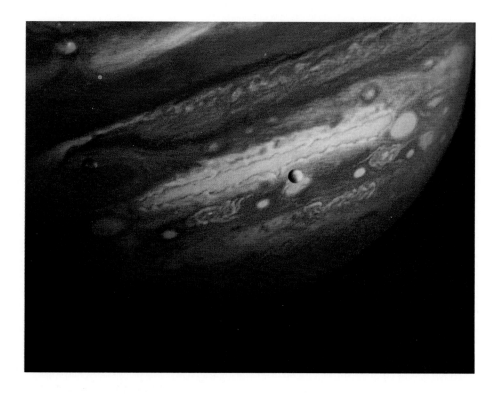

including our own. The reason for this seemingly odd situation is that Jupiter is only slightly more than one-third denser than water, while the Earth and similar planets are about five times denser and, judging by the heat Jupiter emits, it must also have a very high internal temperature. The low density tells us that the planet must be made primarily of hydrogen and helium, the two lightest and simplest of all elements, with only a small admixture of other substances. As a result the planet has an almost solar composition and an almost solar density. The high temperature, on the other hand, guarantees that, in most of the Jovian interior apart from its relatively small rocky core, there are no complicated molecules and compounds whose behavior would be difficult to estimate. Of course, there are some difficulties in trying to understand the behavior of even simple elements at pressures and temperatures that are much higher than those obtainable in laboratories. Nevertheless, these difficulties are gradually being resolved.

Using additional information—such as the rotation period of the planet (9 hours and 55 minutes), the slight flattening at the poles, and the interaction with its numerous satellites—one can deduce a satisfactory model of the planet's

Mechanism of the banding of Jupiter's clouds. Atmospheric convection currents bring the hot, and thus bright, clouds to the surface, where they cool (darken) and return to the deeper layers.

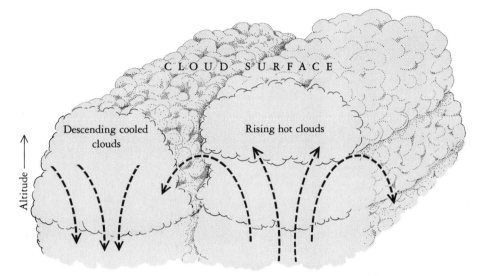

interior. It appears that the core is rocky, liquid, and relatively small, being just over 4 percent of the mass of the planet, although it is 14 times the mass of the Earth. The core is surrounded by a thick layer of fluid hydrogen containing some helium. We expect the deeper part of this layer to be liquid and metallic, because all materials when sufficiently compressed should behave as metals— that is, should have high electrical conductivity—and hydrogen does so at a pressure of about 2 million atmospheres. At distances greater than three-quarters of the way from the center to the outside of the planet the pressures are lower, and hydrogen assumes its usual molecular form. At still greater distance from the center and correspondingly lower pressure, the hydrogen gradually becomes gaseous and, with a small addition of other gases, forms the outer atmosphere of the planet.

Jupiter has such high pressures and temperatures that it has no solid or liquid surface; everything is fluid, made of liquids and gases, without a distinct boundary between the two. What we see from the outside is a dense layer of ammonia crystals and of ammonia compounds at a temperature of about 123° K (−150° C). Occasionally, through a break in them, we have glimpses of deeper layers at a temperature near 300° K (20° C), or close to the average temperature of our own atmosphere. Therefore, it is plausible to assume that below the icy clouds of ammonia droplets there are other clouds such as those of water ice— somewhat like the very fine high-altitude clouds on the Earth—and, still lower, water droplets.

Ever since the highly successful two Pioneer (1973 and 1974) and two Voyager (1979) missions to Jupiter, we have had at our disposal a rich collection of color photographs of the visible Jovian ammonia clouds. The hydrogen and

The Great Red Spot, major atmospheric feature of Jupiter, is about 60,000 kilometers across. Color is computer-enhanced.

helium, being transparent, are not visible. The clouds are various shades of orange and yellow that are presumably caused by an admixture of some other compounds, perhaps containing organic molecules or sulfur or phosphorus. It is these clouds that form the familiar striking pattern of dark and bright bands parallel to the planet's equator. In the bright bands the hot atmospheric gases rise, and in the dark bands the cooled gases descend.

Of particular interest is the enormous Great Red Spot near the equator, which was discovered in 1664 by British scientist Robert Hooke using Galileo's telescope. It is a unique and spectacular oval feature some 40,000 kilometers long (about the length of the Earth's equator) and 13,000 kilometers wide. It is somewhat cooler than the surrounding clouds, and its color presumably indicates the presence of red phosphorus. Dozens of proposals have been made to explain this long-lasting anomaly in the otherwise variable and mobile clouds. These explanations range from assumption that the GRS, as it is usually called, is nothing more than an enormous hurricane that has lasted at least 300 years, to the more exotic one that it is a column of quiescent fluids rising above a solid hydrogen island floating in the liquid deeper layers of the planet. None of the proposals has explained all the many peculiarities of the spot and, in particular, its persistent eastward or westward motions, with respect to other visible features, at a constant speed of about one meter per second. Some of these excursions have continued in the same eastward or westward direction for so many years that, with respect to the other clouds, the GRS has gone all the way around the planet several times. The spot occasionally changes its size and color but shows no southward or northward motion. There is no doubt that within the spot there is a rapid circular motion that carries various small irregularities— huge by terrestrial standards—around it, like pieces of wood in a gigantic whirlpool. Observations show that it takes about a week or two for one of these dots to

The White Oval (just below the GRS) and other ovals.

travel around the GRS and that after one or more such trips they are usually ejected into the adjoining cloud belts.

The actual colors of the Jovian clouds and of their finer features are all somewhere between yellow-orange and orange-red and their motions are difficult to follow. For this reason, most studies of the highly convoluted and complicated motions of the clouds are based on photographs and movies in which the color contrast has been artificially—that is electronically—enhanced by computer.

Composite image of the southern hemisphere of Jupiter made from pictures taken by Voyager. The GRS can be seen in the upper left corner in this orientation, which is directly above the southern pole. Many white ovals, fairly uniformly spaced, can also be seen. The dark irregular form at the center is an artifact of the imaging process.

As a result, some of the images show white, brown, green or blue clouds, which make them resemble non-objective paintings, particularly those by such artists as Kandinsky and Corneille. The quality and resolution of these images is so high that we can distinguish details as small as 30 kilometers across—a truly remarkable achievement. As yet, we have no generally accepted explanation of these turbulent motions, streams, filaments, feathery edges, plumes, delicate curlicues, and reddish or whitish spots. The small dots seem to chase one another and are of rather short duration. One of the larger features, which without artificial color enhancement looks white, is called the White Oval, and is the size of our Moon. It shifted with respect to the GRS by 90 degrees in longitude between March and July, 1979—that is, between Voyager 1 and Voyager 2

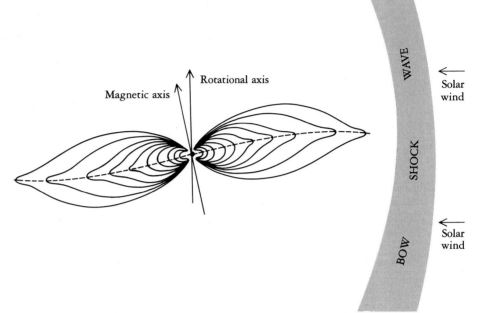

Jupiter's magnetic field. Note that near the planet the field has a shape that is much like the one that normally surrounds a small magnet but is deformed further out.

flybys. Other, usually smaller white ovals and their seemingly regular spacing are clearly visible in the images of the northern and southern polar hemispheres.

Even on the dark, night side of Jupiter, eye-catching features appear—enormous aurorae above the north pole, perhaps bright meteor trails in the atmosphere, and clusters of lightning similar to those we know on Earth, but on a much larger scale.

For a long time before any spacecraft flew by Jupiter it was known that the planet had a magnetic field. The argument was based on the observation of the emanation from Jupiter of strong electromagnetic signals, not unlike radio waves, which we know are usually produced when electrons are trapped by a planet's magnetic field. All four spacecraft penetrated the magnetic field of Jupiter and measured it at various distances from the planet. It appears that, on top of the cloud deck, the field is about eight times stronger than Earth's and that within a few planetary diameters from the surface, its shape is similar to the terrestrial field, resembling that of a bar magnet. Surprising was the find that farther from the planet the field is very distorted and peculiarly elongated. The distortion is caused by a huge ring, or belt, of plasma that is a mixture of electrons and hydrogen ions and that surrounds the planet and co-rotates with it. These particles push the magnetic field outward so that it forms an enormous flat region, the so-called magnetosphere, or magnetodisk. The Jovian field is, however, not free to spread toward the Sun because of the Sun's own magnetic field, which spreads out together with the solar wind. Depending upon the strength of the solar wind, which may vary considerably, the Jovian magnetosphere extends toward the Sun anywhere between 25 and 50 of the planet's diameters—that is, three to six million kilometers. Were this huge disk visible, it would appear bigger in our sky than the disk of the Sun or the Moon. In the direction opposite to the Sun there is no limit to the expansion of the field, and there the magnetosphere, aided by the solar wind, forms what is called a magnetotail, which has been observed as far from Jupiter as Saturn—that is, as far from Jupiter as Jupiter is from the Sun, a truly gigantic feature.

The magnetic axis of the Jovian field, that is the north-south direction, makes an angle of some 10 degrees with the rotation axis of the planet. As a result, the magnetodisk does not lie exactly in the equatorial plane of the planet, and so, as the planet rotates, the disk seems to flop around like an enormous hat brim tilted on the head of a spinning skater. One might think that the powerful Jovian magnetic field and the magnetosphere would keep all the electrons and ions securely trapped in it. In actuality, some of the electrons acquire such high energies that they escape from the flopping and slightly leaky magnetosphere. These high-energy electrons from Jupiter are sprayed throughout the solar system and have been observed near the Earth, identified by their 10-hour period, which is the approximate period of rotation of Jupiter and of its magnetosphere.

As we have noted, the magnetic field of a rotating planet is produced by electric currents generated in its often metallic liquid interior by thermally driven motions. On Jupiter, such a magnetic dynamo operates in the liquid metallic hydrogen layer described above. The Jovian field, whose origin and strength seem to be well understood, is the most powerful field in the solar system apart from the field of the Sun itself.

Next to our Moon, the most famous satellites are the Galilean, the four largest of Jupiter's fourteen: Io, Europa, Ganymede, and Callisto—in Greek mythology, the lovers of Zeus (called Jupiter by the Romans). The discovery by Galileo, in 1610, of these satellites and their motion around the planet played an important role in confirming his ideas about the similar motion of planets around the Sun. Subsequently, the regular and easily observed motion of these satellites permitted Danish astronomer Olaus Römer to determine the velocity of light. He reasoned that the Jovian satellites circle the planet in almost the same plane as the Earth circles the Sun, and during its yearly motion, the Earth varies its distance from Jupiter. The difference between the largest and the smallest distance is the diameter of the Earth's orbit around the Sun. Light from the Jovian satellites, which move rapidly around the planet, reaches us therefore,

An aurora on Jupiter (the bright double streak) and what are probably lightning flashes (the bright spots in the center and lower portions).

sometimes after a short trip and sometimes after a long one. Using precise timing of the moments when the individual satellites disappeared behind Jupiter or emerged on the other side of it, Römer calculated that fifteen minutes is the time necessary for their light to cross the diameter of the orbit of the Earth, a distance of about 300 million kilometers. This information gave him directly the speed of light. His measurements gave a value which was very close to the 300,000 kilometers per second that we use today, a remarkable achievement considering that Römer made his calculation in the year 1675.

Recent detailed observations of the Jovian satellites by Pioneer and Voyager spacecraft brought them again to the forefront of interest, especially if one looks at them as a small version of a planetary system. From previous measurements we

The four large Galilean satellites of Jupiter. These are, clockwise from the upper left, Io, Europa, Ganymede, and Callisto. The latter two are larger than Mercury, and Io and Europa are about the size of our Moon. Image resolution is about 50 kilometers, except for Callisto, where it is about twice that. The sizes of the planets increase in the order Europa, Io, Callisto, and Ganymede.

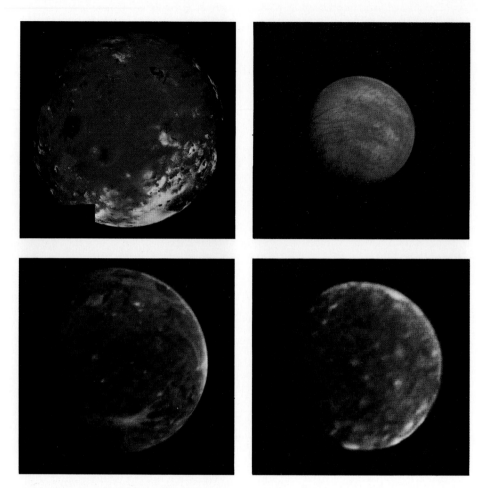

knew that all four of the Galilean satellites are locked in fixed orientation with respect to Jupiter so that, like the Moon, the same side always faces the planet; there is also a difference in coloration of their leading and trailing hemispheres, the cause of which is not well understood. All the satellites except Europa are slightly bigger than our Moon and all have densities comparable to it. Actually, the two satellites farthest from Jupiter are significantly less dense than the other two, suggesting that long ago, when the planet was still very hot, it heated the satellites to high temperatures, driving off volatile elements, which escaped more readily from the nearer, hence hotter, ones, making them denser.

Io, the nearest to Jupiter of the four satellites, is the only body in the solar system besides the Earth on which active volcanos have been directly observed. This volcanism was discovered during a procedure designed to help ascertain the exact location of a spacecraft with respect to a celestial body by obtaining an image showing the body's circumference against the background of the distant stars. The picture of Io taken for this purpose on March 8, 1979, from a distance

Volcanic activity on Io. a *shows a volcanic caldera (light-blue patches), and* b, c, *and* d *show side views of a volcanic plume. The colorful areas around the black craters are sulfur deposits. Color is computer-enhanced.*

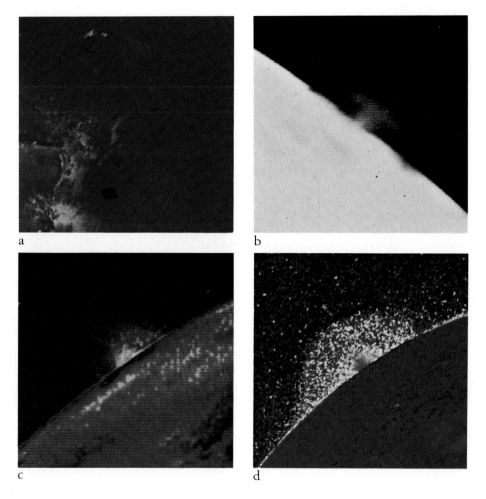

a

b

c

d

of about 4½ million kilometers showed not only a peculiar white blob at the boundary between the dark and the illuminated part of the planet but also a rather faint plume visible against the black sky. Linda Morabito, who made this discovery, and her colleagues at the Jet Propulsion Laboratory, concluded that both features must be signs of powerful volcanic eruptions. The first one was perhaps 300 kilometers high and the second more than 100 kilometers, which is larger than any volcanic eruptions on the Earth. Further studies revealed that Io has nearly a dozen active volcanos, which appear as black spots some 200° C hotter than the surrounding territory and a few tens of kilometers in diameter. A careful study of these volcanos by Susan Kieffer of the U.S. Geological Survey has shown that they are probably filled with molten sulfur and sulfur dioxide and that the plumes are most likely sulfur dioxide in the form of fine particles. This compound decomposes into oxygen and sulfur and the latter gives the satellite its yellow-orange color and, in particular, the many-colored deposits around the volcanos. The reason for the enormous heights of Io's volcanic plumes, as compared to those on the Earth, is, of course, that gravity on Io is five times weaker. The material is ejected at up to 1000 meters per second, three times the velocity of sound on the Earth. The total amount of material ejected in these plumes and surface flows is about 1000 billion tons per year, which has covered the surface of Io to a depth of 100 meters in a million years. Thus, it is hardly surprising that we see no evidence of meteoritic impacts on Io although it has undoubtedly been as heavily bombarded as the other Galilean satellites.

The origin of the energy necessary to eject the enormous amounts of material from Io's volcanos—as unusual as the volcanos themselves—was postulated as part of a prediction by Stanton Peale of the University of California and Pat Cassen and Ray Reynolds of the NASA Ames Research Laboratory that there were volcanos on this satellite. Their reasoning was based on the fact that Jupiter produces tidal deformations in Io that are much bigger than those produced by the Earth on our Moon. These deformations are periodically altered by the other Galilean satellites, especially Europa, with the result that Io is alternately squeezed and stretched. Such huge mechanical deformations result in heating, which keeps Io's interior molten and encourages volcanism. The solar and internal heat explain why, in contrast to the other Galilean satellites, there is no and probably never was any ice on Io. The sulfur dioxide, decomposed by solar radiation into free sulfur and oxygen, forms a thin atmosphere around the satellite. There is also a doughnut-shaped belt or torus along Io's orbit that also contains sulfur and oxygen as well as gaseous forms of elements such as sodium, all of which are ejected from Io or scoured from it by energetic particles in the Jovian magnetosphere that bombard its surface. Trapped in the rapidly rotating Jovian magnetosphere, the particles become heated to some 100,000 K, producing radiation that can be observed from the Earth, especially the yellow sodium light.

Io demonstrates other curiosities. The satellite is a good conductor of electricity and, when it moves rapidly through the Jovian magnetic field, acquires electrical potential across its diameter that produces an electric current estimated at 10 million amperes. This current flows along a line, or magnetic flux tube, between Io and the upper atmosphere of Jupiter. As the planet rotates, the foot of the flux tube traces a path in the atmosphere, producing powerful bursts of electromagnetic radiation easily observable from the Earth. The existence of this flux tube was confirmed by spacecraft flying close to it. Similarly the bright aurorae in the polar regions of Jupiter are caused by charged particles escaping

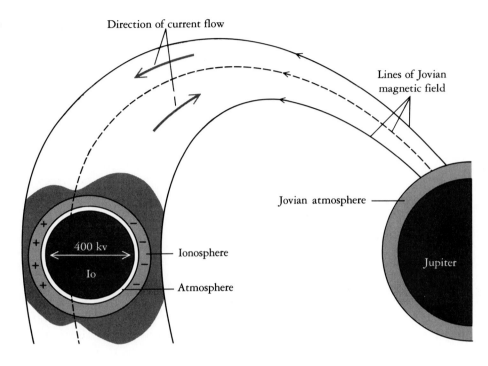

Positive and negative charges produced on Io as it moves through the Jovian magnetic field generate a powerful electrical current between it and the planet. The existence of this flux tube was confirmed by direct observations from spacecraft. (Figure is not drawn to scale.)

Direction of current flow

Lines of Jovian magnetic field

Jovian atmosphere

Ionosphere

Atmosphere

Jupiter

400 kv

Io

from Io's torus. All in all, Io appears to be unique among the satellites of the solar system.

The remaining three Galilean satellites are less unusual than Io, but each has its own puzzles. Europa is a bit smaller in diameter than our Moon, and its density indicates that most of its interior is probably made of rocks, though the outer 100 kilometers or so are mostly ice. The ice reflects the incoming solar light and the satellite is very bright. Its surface is covered with a network of fairly straight dark lines, some of which are up to 3000 kilometers long and 70 kilometers wide. So far, no successful interpretation has been made of these mysterious lines or of bright ridges a few hundred meters high, another surface feature of the satellite. Some astronomers think that below the mostly icy crust there is liquid water, not unlike parts of our Arctic Ocean, and that the water permits considerable motion of the crust, leading to cracks and ridges. The absence of obvious impact craters on Europa is either the result of the softness of ice and its deformability or of the fact that the ice surface was formed after the primordial period of bombardments.

The next Galilean satellite, Ganymede, is the largest in the solar system; indeed it is larger than the planet Mercury. Its density is so low that the satellite

Network of lines up to 3000 kilometers long on the ice-covered surface of Europa.

must be about half rock and half ice or water. Ganymede has, in contrast to the fairly uniform Europa, a highly diversified surface. Some areas, comparable in size to the United States, are dark; others show valleys and mountains some 1000 meters high, evidence of an active geological history. In fact, this satellite seems to show evidence of plate tectonics, not unlike that on the Earth. The many impact craters indicate that the satellite was once heavily bombarded, and the difference in density of distribution of these craters in various parts of Ganymede suggests considerable differences in local ages of the crust. The fact that some of the craters are nearly white may be evidence that the crust is rocky and thin and that the underlying ice becomes exposed on impact. There are also areas full of parallel grooves and ridges 10 to 15 kilometers wide and up to 1 kilometer high resembling the well-known parallel mountainous topography near Harrisburg and Allentown in Pennsylvania.

Callisto, most distant from Jupiter of the Galilean satellites, is the size of Mercury. Early in its history it must have been heavily bombarded, and the many surviving impact craters indicate that, in contrast to Ganymede, there was

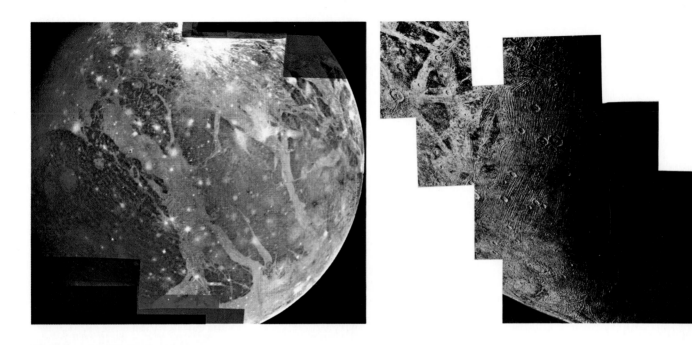

The largest satellite in the solar system, Ganymede shows a highly varied surface. The white craters suggest that impacts on the planet broke through its crust, releasing water, which froze rapidly. At right, a closer view of part of the surface shows the conspicuous parallel grooving, which is composed of mountain ridges 10–15 kilometers apart.

no internal geologic activity that would have obliterated old craters. Callisto is thought to have a surface of ice and rocks with the result that the easily flowing ice erased all impact craters larger than about 200 kilometers. The thickness of the ice or water layer under the rock-strewn crust is estimated to be three times that on Ganymede. Callisto's density is, accordingly, lower than that of any of its companions. It has no significant mountains or valleys, and a typical range of surface temperature during the day is 75 K (−200° C) to 155 K (−120° C), which is drastic.

A surprise of the Voyager missions to Jupiter in March and July of 1979 was the confirmation of the suspected existence of rings. These rings are much nearer to the planet than any of its major satellites and much smaller than the well-known rings of Saturn. They consist of an outer bright part about 800 kilometers wide, an adjoining fainter ring 6000 kilometers wide, and a still fainter ring that seems to extend all the way to the planet's atmosphere. The distance from the outside edge of the ring to the visible cloud tops of the planet is less than half the planet's diameter. The rings are so faint and thin and so close to this very bright

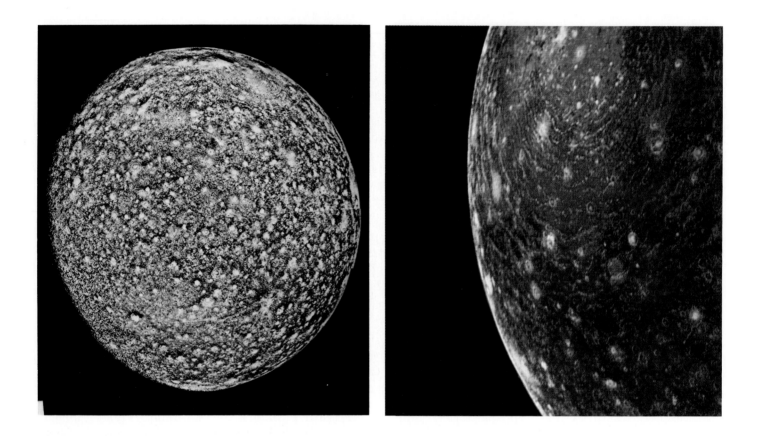

Two views of Jovian satellite Callisto, the most heavily cratered in the solar system. Close-up view at right shows a large impact basin, similar to the Caloris basin on Mercury. It is the first found in the Jovian system.

planet that they were not detected before Voyager; since then, aware of what we were looking for, they have been observed, albeit with difficulty, from the Earth.

Before the discovery of Jupiter's ring system, only two others were known—those of Saturn and the recently discovered rings of Uranus. The presence of rings around a planet was explained in 1850 by French mathematician Edouard Roche, who concluded that tides produced by a planet's gravity can break up a satellite or a passing body if it comes close enough to the planet. We know that the Earth's gravitational force produces stresses and tensions in the Moon that would have been great enough to have broken it up were it much nearer the Earth. Roche has shown that if a passing body comes nearer the surface of its parent planet than about 1.25 times the planet's diameter, it will break up. The precise limit depends, of course, on the strength, density, and shape of the satellite. If the satellite is small, it will not break up.

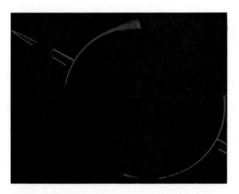

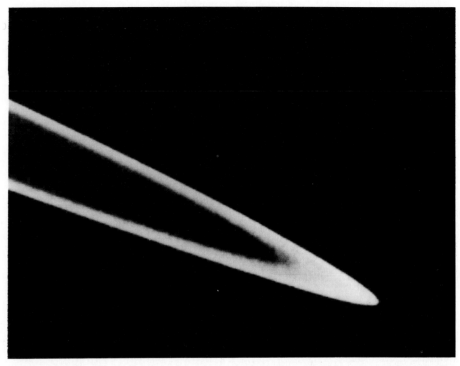

Above, the night side of Jupiter with the planet's atmosphere back-lighted by sunlight. Also shown are parts of the rings and, at right, a detail of the ring showing the thinness of the disk, which probably extends to the planet's surface.

Most of the planetary rings we know lie at distances smaller than the Roche limit. When a break-up of a satellite occurs the initially quite large fragments will eventually, as a result of collisions among themselves, become smaller and spread into a ring containing a range of sizes of particles. Roche's conclusions imply also that small fragments orbiting a planet at distances smaller than the Roche limit will not adhere to each other and thus will not grow to form a larger body. Interestingly enough, the particles in Jupiter's ring seem to be so fine that a large part of them is probably produced by the volcanic ejecta from Io. These high-velocity ejecta escape the gravitational pull of Io, bombard a small rocky satellite at the outer edge of the ring, and break tiny fragments from it. The biggest fragments form the brightest outer ring; the finest drift down toward the planet, forming the faint halo. Thus, the ring gains particles from its small satellite and loses them to the planet itself. We know that the ring particles must be rocky because various ices would be unstable that close to the hot planet.

Between the four big Galilean satellites and the intriguing ring close to the planet, Jupiter has three very small, newly discovered satellites (about which we know very little) and one satellite, Amalthea, discovered in 1892 and named after the mythological nurse of Zeus-Jupiter. This satellite is only 265 kilometers long and 150 kilometers wide and, as would be expected, its long axis points toward the planet. Amalthea is reddish dark and is heavily bombarded by particles in the Jovian magnetosphere and by those escaping from Io. Besides two very small satellites near the rings and one between Amalthea and Io, Jupiter has, farther out, beyond Callisto, eight other outer satellites about which we know very little except that, in contrast to the Galilean satellites, they did not form from the same initial gaseous cloud as Jupiter, but are captured bodies, probably asteroids or comets. There are two reasons for this conclusion: First, their orbits are highly eccentric and do not lie in the equatorial plane of the planet as do the inner satellites; and, second, four revolve in one direction, and four in the opposite—a truly irregular bunch of accidental hangers-on. Spectroscopic studies show that the two groups also have different chemistry.

SATURN

Saturn with its rings, which can be seen even with a pair of good field glasses, is undoubtedly the most spectacular of planets. It is also a giant,—though its diameter is about nine-tenths that of Jupiter and its mass is about one-third.

Saturn resembles Jupiter in many ways but also differs from it significantly. Its density is about half that of Jupiter, less than three-fourths that of water, so that if there were a sea large enough to contain it, Saturn would float in it. In fact, it is the least dense of all the planets, with the possible exception of Pluto, whose density is still poorly known. Like Jupiter, Saturn emits about twice as much heat as it gets from the Sun. An important difference, however, is that while the internal heat of Jupiter is easily accounted for by the heat accumulated during its formation, Saturn has not retained enough of that heat to account for its present level of radiation. The origin of its extra heat was pointed out in 1967 by the author; it is as follows: Below certain temperatures and pressures liquid metallic hydrogen and helium are not completely soluble in each other—that is, condensed hydrogen dissolves only a little helium and helium dissolves only a little hydrogen. At very high temperatures and pressures, on the other hand, the two substances are totally soluble in each other—like, for instance, water and alcohol. On Jupiter, conditions are such that the two elements are probably completely dissolved in each other, but, according to David Stevenson of the California Institute of Technology, this is not the situation on Saturn. As the planet very slowly cools, the amount of helium that can be dissolved in hydrogen decreases, and the resulting excess helium appears as small droplets, which, being heavier, drift downward. This situation is a familiar phenomenon in cooking: Sugar or salt dissolve easily in a glass of hot water, but as the water cools,

Detail of the Cassini division (limits indicated by arrows), showing that the division, once thought to be empty, contains many ringlets.

excess sugar or salt appear as small crystals on the bottom of the glass. As the helium droplets move toward the center of the planet they release a certain amount of gravitational energy. It is this extra energy that is the source of Saturn's additional internal heat.

According to Stevenson the drift of helium-rich metallic fluid to the deeper parts of Saturn has two further consequences, which confirm the presence of separate helium-rich and helium-poor layers in the planet's interior. First, the magnetic field, generated only in the deep metallic layer, is much weaker than it would be if the separation between liquid helium and hydrogen did not exist and the field were produced in the helium-poor layer as well. Second, direct measurement by a spacecraft showed the concentration of helium in the atmosphere of Saturn to be 11 percent, much lower than Jupiter's 19 percent—again in accord with the idea that, on Saturn, most of the helium has drifted to much deeper layers than on Jupiter.

Bright and dark zones in Saturn's clouds, similar to but less distinct than those of Jupiter, and, below, a detail showing the waves and eddies making up the bands. Color is computer-enhanced.

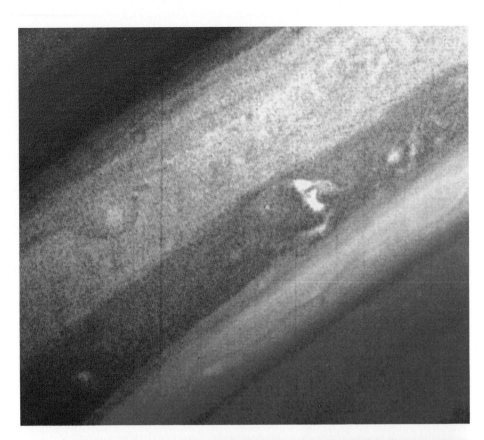

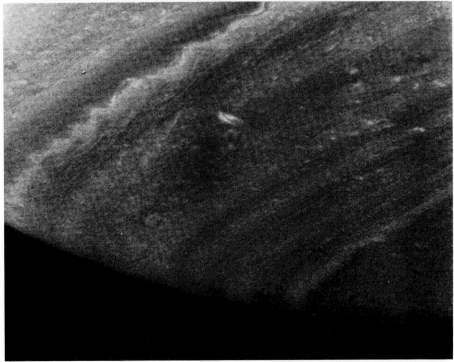

Convection cells and a ribbon-like streamer in Saturn's cloud system.

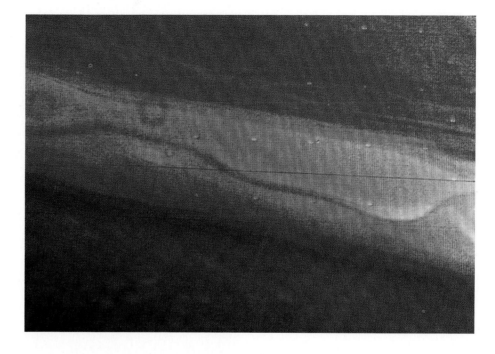

For a long time there was considerable uncertainty about the period of rotation of Saturn, or rather of its visible clouds. The equatorial belt of clouds has a speed four times that of Jupiter's and moves around the planet some 10 percent faster than the regions near the poles. A method of estimating the rotational period of the planet itself, rather than of its clouds, is to measure periodic changes in the intensity of radiation emanating from the planet's magnetosphere. This procedure worked very well with Jupiter but could not be applied to Saturn because no such radiation could be detected from the Earth. In 1979 both Voyager spacecraft were able to observe weak periodic bursts of radiation near the poles, which permitted determination of the rotation period of the body of the planet at 10 hours, 39 minutes. Earlier detection, it turned out, had been thwarted by the planet's rings, which, as James Van Allen of the University of Iowa indicated, captured particles that otherwise would be trapped by the planet's magnetic field and would have radiated toward the Earth. The particles trapped by the weak field outside the Saturnian rings could not produce an effect strong enough to be measured from the Earth.

Like Jupiter, Saturn shows bright and dark zones parallel to the equator but, probably because of thicker ammonia clouds, the contrast between them is much lower and the colors much blander than on Jupiter. Electronic enhancement,

A portion of the rings of Saturn and their shadow on the surface of the planet. Visible are the outer part of the A-ring, the thin Encke division, the inner (wider) part of the A-ring, the wide Cassini division, the very bright B-ring, and the much fainter C-ring. The reddish-white irregularity on the edge of Saturn's disk is a photographic artifact.

however, delineated details, including reddish and pale-orange spots. One reddish spot is similar to the Great Red Spot of Jupiter, though a seventh the size. Voyager 2 data seem to indicate that this spot, like the GRS, is an enormous whirlpool of clouds. Since these spots on Saturn are not easily observable from the Earth, we know next to nothing about their motion with respect to the surrounding clouds. There are, however, features that do not seem to exist on Jupiter, such as long dark ribbons or streams in the clouds tens of thousands of kilometers long.

In 1610, Galileo observed through his telescope anomalous bulges on both sides of Saturn and thought, at first, that what he saw were two huge satellites or perhaps a triple planet. Later, to his great surprise, the bulges disappeared from view. Then in 1655, Dutch astronomer Christiaan Huygens gave the proper interpretation by pointing out that what Galileo had observed was a flat thin ring, or disk, seen sometimes from above and sometimes from below and that periodically, when seen edge-on, would almost disappear from sight. It was not until 1859, however, that the great Scottish physicist James Clerk Maxwell theorized that the ring must be composed of small particles each in its own orbit around the planet, because a solid disk or plate would have been broken up by tidal effects produced by the parent planet. His argument was similar to the reasoning of Roche against the existence of large satellites very close to a planet. Some six years later, American astronomer James Keeler further confirmed Maxwell's theory by showing, with spectroscopic methods, that a portion of the ring nearer the planet revolves around it faster than the rest and, thus, the ring could not be solid.

Saturn's rings are undoubtedly the best-known feature of this planet, but until the results of the Pioneer and Voyager flybys between 1973 and 1981 became available and were properly interpreted, our knowledge of the rings was limited. Astronomers used to distinguish three broad rings: the outer, fairly bright A-ring, reaching some 135,000 kilometers from the center of the planet, with the 200-kilometer-wide Encke division in it; then a much brighter B-ring, separated from the A-ring by the 3500-kilometer wide Cassini division; and finally, the much fainter C-ring, which extends to 73,000 kilometers from the center of Saturn. The existence of still fainter rings—one outside Ring A, and a D-ring between the C-ring and the planet—had also often been suggested. The brightness of the rings and more detailed optical studies confirmed the conclusion that the tiny particles were made of, or are covered with, ice, which can well exist in these surroundings. The two divisions, Encke and Cassini, were thought to be entirely empty. It was argued that, among the ten-or-so satellites of Saturn known at the time, there were a few massive enough or near enough to the rings to perturb the motion of ring particles. Such a perturbation occurs when there is a resonance between the periods of revolution of a particle in the ring and of a satellite—that is, when the relation between the two is a simple fraction, such as $\frac{1}{2}$ or $\frac{2}{5}$. Under these circumstances, small particles would be removed from their orbits, leaving a gap. The Encke and Cassini gaps occur where there are such resonances with the nearest two satellites, Enceladus and Mimas.

This fairly simple picture changed greatly when the amazing images obtained by Voyager 1 and especially Voyager 2 became available. They showed that the A-ring has other divisions besides Encke; that neither the Encke nor the Cassini division is empty, each containing several rather regularly spaced narrow faint rings; that all the rings consist of hundreds, if not thousands, of narrow "ringlets," which make the whole system look like a giant phonograph record; and finally, that there is a faint D-ring, composed of several narrow ringlets located between the C-ring and the planet. The fact that the Cassini division turned out not to be empty argues for caution in the planning of spacecraft missions: During such a planning session the question was taken up of whether one of the Pioneer spacecraft to Saturn should pass fairly far from the A-ring or close to it

A segment of the whole of Saturn's ring system. The outermost ring (F-ring) and one of its shepherding satellites are barely visible at the upper left corner, and the Cassini division can be seen to consist of four small ringlets. Though not of the highest resolution, this picture shows hundreds of ringlets; at higher resolution the number rises to the thousands. The contrast and distinctness of the rings diminishes where the disk of the planet appears behind them.

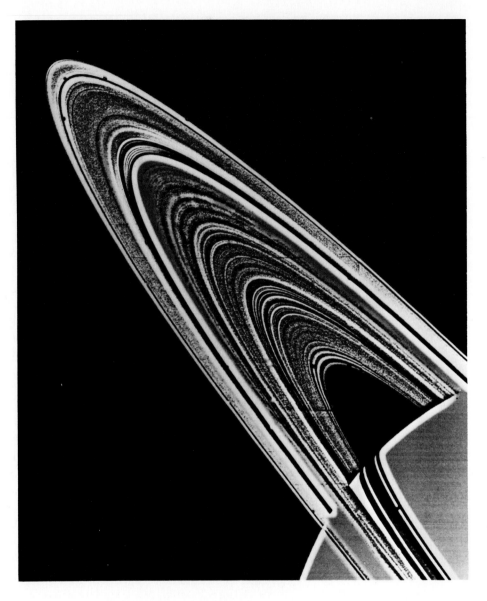

or, in fact, venture through the Cassini division. The author favored the last option, but others prevailed—fortunately, for the collision with even very small particles in the division would have destroyed the spacecraft. Transmission of radio signals from the various spacecraft through the rings indicated that the average size of the particles varies between 2 and 10 meters as one goes from the D- to the A-ring. In the divisions, the particles are fewer, of two sizes, and generally smaller than in the rings.

Entirely new rings were also discovered beyond the A-ring: a faint, narrow F-ring separated from the A-ring by a 4000-kilometer-wide gap called the Pioneer gap, a difficult to observe G-ring, and a 90,000-kilometer-wide E-ring, which seems to surround the orbit of the satellite Enceladus. The new F-ring is

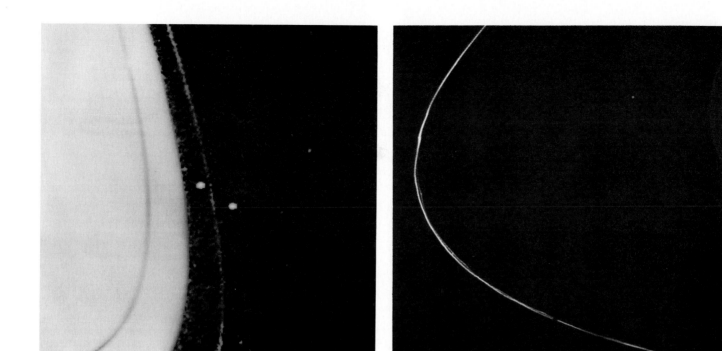

At left, the F-ring with its two shepherding satellites, which keep it from dissipating. In this view they are about 1800 kilometers apart. Within about 2 hours the inner moon passed the outer. At right, in another view, the F-ring can be seen to have a braided structure.

not smoothly circular but consists of two or three somewhat wavy strands. By way of explanation of this the author has suggested that those particles of the ring that are unusually small—perhaps one-thousandth to one-hundredth of a centimeter in diameter—are electrically charged by incident solar radiation and deflected by the Saturnian magnetic field, which results in the observed waviness. The larger particles of this ring are not as easily deflected. Another explanation is based on periodic gravitational influence by two small satellites, each about one hundred kilometers in diameter, one of which has an orbit in the Pioneer gap and the other outside the F-ring. The two satellites, Numbers 1980S26 and 1980S27, because of their controlling relationship to the particles, are called "shepherding" satellites. An electrical charging effect has been proposed to explain the puzzling radial streaks or spokes in the B-ring, which are dark in reflected light but bright in transmitted light. The positions of these spokes appear to be closely coupled to the planet's magnetic field and also to the electrical discharges, not unlike lightning, which have been observed by Voyagers 1 and 2. These lightning bolts are 10 to 100,000 times more powerful than those we see on the Earth.

Several proposals were made to account for the reason all rings seem to be divided into thousands of fine ringlets. There are too many to invoke the role of resonances with the major satellites, as was done with the Encke and Cassini divisions. One suggestion ascribes each ringlet to a small satellite that is contained somewhere in it and is the source of particles broken off by bombardment. Another suggestion is that there is a small satellite between each two ringlets that acts not unlike the shepherding satellites of the F-ring. So far, however, the presence of such tiny bodies has not been confirmed. Clearly, the new world of the Saturnian rings so far revealed is astounding, and the data collected by the spacecraft have only begun to be analyzed systematically. Questions abound about newly discovered phenomena such as the presence of single ringlets, which are not circular but slightly elliptical; the size distribution of ring particles; the presence of dark and bright longitudinal streaks, which used to be interpreted as the result of wakes formed among the finer particles by a big one; and the bright knots and kinks in the F-ring. It is thought that the formation of the Sun or of the planets themselves from the solar nebula or of their satellites from the planetary nebulae may have been preceded by the formation of rings, and thus the issues raised by the new observations may be of truly basic importance to our understanding of the history and evolution of the whole solar system.

Until the spacecraft missions we could see, in order of increasing orbital paths, nine satellites of Saturn: Mimas, Enceladus, Tethys, Dione, Rhea, Titan, Hyperion, Iapetus, and Phoebe. The larger ones, except for Titan, are referred to as icy satellites because of the composition of their surfaces. The last three are very likely captured bodies. In fact, Phoebe orbits the planet in a direction opposite that of the other satellites. The spacecraft images and very advanced telescopic observations added eight small satellites—usually called "rocks"—most of them icy and located between the outer edge of the A-ring and the orbit of Mimas. Some of the new satellites offer curiosities: For instance, the distance of Satellite 1980S1 from the planet is 50 kilometers greater than that of Satellite 1980S3, and thus moves slower around the planet than the other. The diameters of the two satellites are about 100 kilometers, twice the difference in their orbits, and it would seem that these almost co-orbital satellites are bound to collide. Detailed calculations show, however, that when the two satellites come very near each other, they not only do not collide but interchange their orbits, the outer moving to the smaller orbit and the inner one to the larger. This curious do-si-do occurs every few years and has undoubtedly occurred billions of times in the past. The satellite that, at present, is farther out looks like a broken, highly irregular peanut. Both satellites may be fragments of a once much larger body.

The nine larger satellites of Saturn are heavily cratered. Mimas has one 130-kilometer-wide crater, about one-quarter the satellite's diameter, but otherwise looks like a poorly made golf ball. Calculations show that, had the body that formed this large crater been any bigger, it would have shattered Mimas. Grooves and ridges clearly associated with the crater are in aggreement with this conclusion. Enceladus may be the source of the material in the wide E-ring because its orbit coincides with the brightest part of the ring. According to A.F. Cook of the Smithsonian Center for Astrophysics and R. Terrile, of the Jet

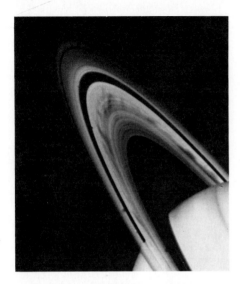

Spoke-like forms in Saturn's broad B ring. These rotate with the motion of the particles making up the rings and the pattern changes continuously.

Artist's conception of a Saturnian ring from within it, with a portion of the planet looming in the background.

a

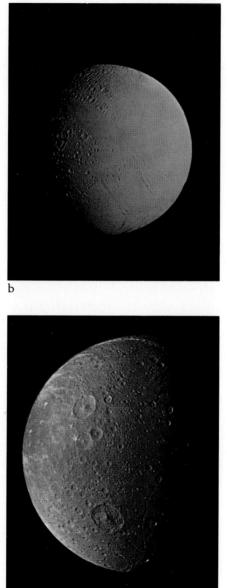

b

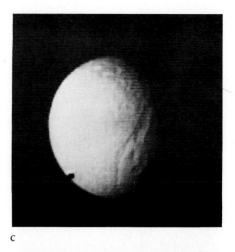

c

f

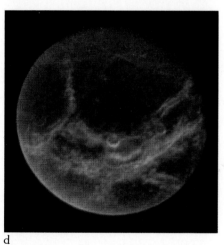

d

g

e

Seven satellites of Saturn in the order of their distance from the planet: a, Mimas, heavily cratered, is about 385 kilometers in diameter; b, Enceladus, with an icy surface, has large areas uncratered and no craters larger than 35 kilometers in diameter; c, Tethys, heavily cratered, shows a valley about 750 kilometers long, while its own diameter is only about 1000 kilometers; d, Dione, shows bright streaks, some of which seem to be grooves resulting from fracturing; e, Rhea, substantially cratered, has ridges and grooves that resemble those on our Moon and on Mercury; f, Titan, enveloped in a thick haze, has a substantial atmosphere that may be denser than Earth's; and g, Iapetus, has a trailing hemisphere (to the right) four to five times brighter than its leading hemisphere. The color of Titan is computer-enhanced.

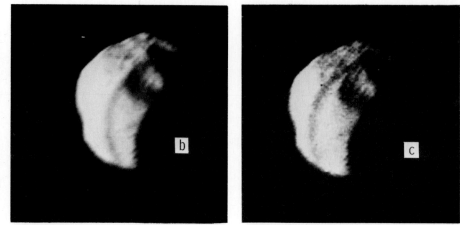

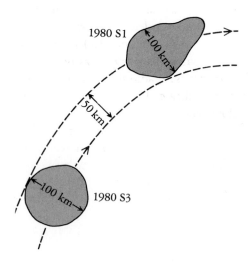

Below, the paths of the co-orbiting satellites of Saturn 1980S1 and 1980S3, the inner of which has the higher velocity. At their closest approach to each other they interchange positions. Above, a series of photographs clearly show the shadow of the A-ring moving across 1980S3.

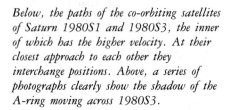

1980 S1

100 km

50 km

100 km   1980 S3

Propulsion Laboratory, Enceladus may have a substantial layer of liquid water under a thin crust of ice, so that, when meteorites collide with the satellite, the water escapes and freezes and the pieces of ice become part of the ring. The author suggested instead that, even if there were no liquid water under the icy crust of the extremely cold satellite, an impact by a fast meteorite must lead to ejection of liquid water droplets and of water vapor. Both water and vapor would freeze and produce small particles, which would then supply the E-ring. In this way, the steady loss of particles that are lost from the E-ring is compensated. The surface of Enceladus does not appear to be very old, a fact that agrees with the idea that it may be periodically covered with fresh ice, perhaps formed by the water exuding from impact punctures. This process would be analogous to the formation of the huge mare on the Moon but, while there it was lava or magma that flowed through a punctured crust, here it would be water or water-rich minerals. There has been speculation that the water layer beneath the icy crusts of Europa and Enceladus may harbor some primitive form of life.

The next satellite, Tethys, is also heavily cratered and has an enormous long valley or rift of unknown origin that reaches three-quarters of the way around the satellite. Dione shows areas of marked brightness contrast and streaks and may have undergone considerable geological transformation. Images of Rhea have been obtained from an altitude of only 59,000 kilometers that show details as small as 1 or 2 kilometers. Its surface is heavily cratered and resembles somewhat that of the Moon or Mercury. Among the other major icy satellites it is worth mentioning Iapetus, which is curious in that the hemisphere oriented in the direction of its motion is five times darker than the opposite hemisphere. The bright trailing hemisphere is most likely ice; the dark side is reddish as in the Jovian satellite Callisto. The reason for the contrast is not known, although some scientists suggest that the dark matter may be organic. The rather distant satellite Hyperion has an irregular flattened shape, not unlike a hamburger patty, and curiously enough, its longest diameter does not seem to be pointing at Saturn. It may be the largest fragment of a bigger satellite that broke apart in a collision. Finally, there is the 200-kilometer diameter Phoebe, farthest known

The satellite Mimas in a view showing a relatively huge (135 kilometers in diameter) crater on this 400-kilometer-diameter satellite. Had the impact that produced it been slightly greater it would almost certainly have shattered the satellite.

satellite of Saturn. It is about four times as far from the planet as Iapetus, and is almost certainly a captured body—probably an asteroid, because it is too dark to be icy, its orbit is highly inclined, and it revolves around the planet in a direction opposite that of all other satellites. The stable of Saturnian satellites and rings is indeed rich, diversified, and puzzling.

Nothing, however, compares with the mystery of Titan, the largest of Saturn's satellites and the only one in the solar system that has a substantial atmosphere. Its orbit lies between those of Iapetus and Hyperion. Titan's diameter is 5 120 kilometers, which makes it slightly smaller than the Jovian satellite Ganymede but larger than the planet Mercury, which in turn is larger than our Moon. The great interest in this satellite arose in 1944 when Gerald P. Kuiper of the University of Arizona discovered evidence of methane gas, indicating, for the first time, the presence of an atmosphere. The very existence of an atmosphere raised many speculations, sometimes quite fantastic, that Titan's surface temperature might permit the development of life. Voyager results seem to have altered these ideas considerably and dampened the hopes of finding life. The best images obtained by this spacecraft show a solid layer of clouds, optically impenetrable. An orbiter or probe using an infrared radiation detector or radar could give an image of the topography as was done with Venus. About 100 kilometers above the cloud layer are one or two fairly transparent hazy layers easily seen in images of the edge of the Titan's disk. Measurements of radio signals transmitted by Voyager 2 as it passed behind the satellite indicate that the solid surface is about 240 kilometers below the visible cloud layer.

Titan's clouds are reddish, which is presumably due to organic molecules derived from the methane gas in the atmosphere. Many of these molecules have been identified. The incoming solar radiation, cosmic rays, and particles of the magnetosphere, which flow by rapidly, break up the molecules of methane into simple organic building blocks, from which more complicated, perhaps biologically significant, molecules may be formed. As reported by Voyager 2, some of these molecules form a genuine smog in the clouds of the satellite. Besides methane, which is observable from the Earth, there is nitrogen on Titan, as predicted by Donald Hunten of the University of Arizona, as well as some carbon monoxide and carbon dioxide. The predominant constituent of Titan's atmosphere is, however, nitrogen, which on the surface of the satellite may reach a pressure one and a half times as high as that in which we live on the Earth. Initially, there seemed a chance that the temperature on the surface was low enough to form liquid nitrogen pools and lakes, but we now know that it is around 94 K ($-179°$ C), which may permit the formation of liquid methane or even propane, but not liquid nitrogen. The chances of the existence of life at these temperatures and under these conditions is essentially nil.

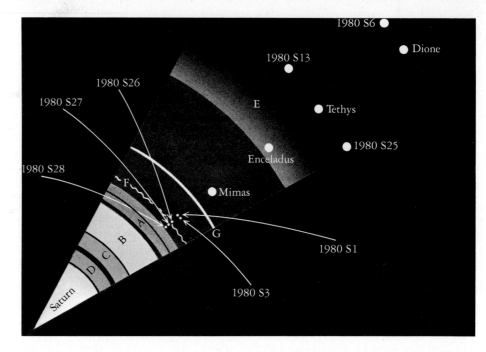

The neighborhood of Saturn. Newly discovered satellites are identified by number rather than name. Note F ring-shepherding satellites S26 and S27, the co-orbital satellites S1 and S3, and the wide E ring, which embraces the orbit of the satellite Enceladus.

Although Titan does not have a magnetic field of its own, it perturbs significantly part of the much-faster-rotating magnetosphere of Saturn. As a result, Titan's substantial magnetotail precedes it along its orbit.

Ever since the detailed spacecraft images of the inner and outer solar system became available we have known that all the terrestrial planets and essentially all satellites of the giant planets show evidence of heavy impact cratering at some earlier period. Were all the craters made by similar bodies and where did these come from? Not many aspects of this question can be answered in a definite manner, although studies of the relationship between the number and the sizes of craters have been very useful. There seem to be two main possible sources of the bombarding material: planetesimals surviving from the early period of the evolution of the solar nebula and formation of the planets, and the debris left by comets. The belt of asteroids located between Mars and Jupiter is the most obvious and huge reservoir of such planetesimals but, as pointed out by Eugene Shoemaker of the U.S. Geological Survey, some of the planetesimals must have been thrown to the outer parts of the solar system by gravitational interaction with the giant planets. Comets, as we know, shed their debris constantly throughout the solar system. There seems evidence that most of the cratering in the inner solar system was caused by material from the asteroid belt. On the other hand, the satellites of the giant planets were probably bombarded mainly by material from the comets, although some huge craters, such as those on Mimas and Tethys, suggest planetesimals. The matter is still under intense investigation.

CHAPTER 5   THE COLD PLANETS AND THE SMALL BODIES

CHAPTER 5   # THE COLD PLANETS AND THE SMALL BODIES

Uranus and Neptune are the most distant large objects in the solar system, and both clearly belong to the regular family of planets, although the mechanism of their formation is not well understood. There is, however, an enormous number of much smaller bodies that are not planets, and there is Pluto, which though usually counted as a planet, is very small and seems to have a dubious pedigree.

URANUS AND NEPTUNE

Our knowledge of the planets more distant than Saturn is indeed small. Uranus is about twice, Neptune three times, and Pluto, on the average, four times as far from us as Saturn, although its orbit is so elongated that occasionally it comes nearer to the Sun than Neptune. Direct observations are much more difficult and, so far, no spacecraft has flown by these planets and reported pertinent data. To be sure, Voyager 2 is on its way to Uranus and to Neptune but, even assuming everything goes well, it will not reach them until 1986 and 1989. Without detailed images of these two planets we must content ourselves with what little we can see from the Earth or learn from instruments orbiting the Earth above its obstructing atmosphere.

Long ago, astronomers concluded that Uranus and Neptune are even more similar than Jupiter and Saturn. Their masses and sizes do not differ by more than about one tenth, their densities differ only by a third.

Preceding page: Pluto's satellite Charon as seen from the surface of the planet in an artist's conception.

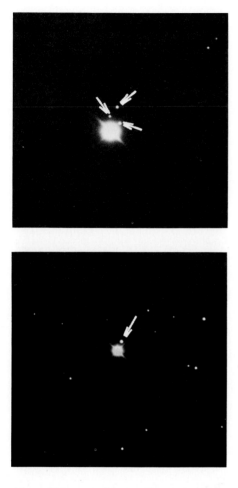

Above, Uranus and its satellites (arrows). The cross-like streaks are caused by supports of the mirrors in the telescope through which the photograph was taken. Below, Neptune and one of its satellites, Triton (arrow).

Most unusual about Uranus is its orientation relative to the Sun, which gives it the appearance of lying on its side, its equator being a plane almost perpendicular to that of the planet's orbit around the Sun. This is in sharp contrast to the other planets, which have their equators tilted to the plane of their orbits by less than 30 degrees and, in some instances, by less than 3 degrees. Uranus takes 84 years to travel around the Sun. During a quarter of that period—that is, for about 20 years—one hemisphere of the planet containing one pole faces the Sun and receives all the solar heat; then, for each subsequent 20 years, all the heat is directed, respectively, at the equator, the hemisphere with the other pole, and then again the equator, and so on. Thus, the poles are alternately hotter and then colder than the equatorial regions. There has been much speculation as to how this peculiar situation would affect the seasons of Uranus. If Uranus' atmosphere is like the Earth's, then the differences between the temperature at the poles and equator are enormous. On the other hand, if the Uranian atmosphere is thick, so that it can store heat for 20 years or more, then the seasonal contrasts are not so great. Preliminary calculations suggest that the atmosphere may store heat for up to 600 years.

How did Uranus acquire this unique orientation of its axis? An important consideration is that the *whole* Uranian system lies on its side: The nine rings and five satellites also lie in a plane that is perpendicular to the plane of the rest of the solar system implying that the anomalous orientation dates back to the very early stages of the formation of the planet itself, rather than to the formation of the solar system as a whole. One of the most successful proposals concerning the unusual orientation of Uranus has been made by Richard Greenberg, of the Planetary Science Institute in Tucson. He assumed that initially Uranus, like the other planets, had its rotational axis perpendicular to the plane of its orbit around the Sun and had an equatorial bulge. Then, very early in its history, Uranus captured a stray body that had a small mass compared to that of the planet. This captured satellite orbited Uranus in a direction opposite to the rotation of the planet at a distance more than 12 times the planet's diameter in a plane that was highly inclined. Tidal interaction gradually brought the satellite nearer to the planet, and the progressively stronger interaction with the equatorial bulge pulled the plane of the orbit of the satellite nearer to the equatorial plane of the planet and, at the same time, gradually tilted the planet's axis away from its original orientation. This process stopped when the satellite came so close to the planet that gravitational forces broke it into several pieces that eventually formed at least three of the present satellites. The planet and its satellites were thus left with a high tilt; whether this is the only possible explanation of the unusual present configuration of the Uranian system is not certain. What is important, however, is that a relatively simple dynamic mechanism can

Uranus, photographed from an airborne telescope about 25 kilometers above the Earth. Under these conditions features as small as 2000 kilometers across would have been detectable.

account for the observations. A corroboration of this tilting mechanism is the fact that Uranus seems to have *no* irregular satellites—that is, satellites that were captured in very elongated and highly inclined orbits. Oberon, its most distant known satellite, orbits Uranus at a distance of twelve diameters of the planet, while Jupiter, Saturn, and Neptune have many regular satellites at considerably greater distances, and captured satellites even further out.

Though early observations suggested a rotation period for Uranus of 10 hours and 45 minutes—which is not very different from the rotation periods of Jupiter or Saturn—later, more precise data, based on different observational techniques, gave values close to either 13 or 24 hours. Usually it is possible to make a good estimate of the rotation period of a planet by timing the reappearance of some distinctive spot near the planet's equator. Since Uranus' axis of rotation at present points almost directly at the Sun and thus at the Earth, the motion of such spots cannot easily be followed, spots at higher latitudes, near the poles, being often too faint to be observed with precision. There are other ways of studying planetary rotation, but they are much less direct and the results less certain. The rotation period of Neptune, now somewhat better known, is between 16 and 20 hours.

The rotational period of a planet is important not only *per se,* but also because it is part of the basic equation that tells us about the planet's interior. Of similar significance is the planet's oblateness, the degree to which it is compressed along the rotational axis. Oblateness is difficult to measure optically if the polar axis is directed at the observer—the present situation of Uranus; it is also difficult to measure in Neptune's case because of the planet's distance. The third quantity crucial for constructing a model of a planet's interior is the shape of its gravitational field. The field of an exactly spherical planet would also be spherical— that is, gravity would decrease with distance from the planet's center at the same rate in all directions. All the outer planets have an equatorial bulge, and the distribution of their inner layers is not exactly spherical. For this reason, it is important to know the deviation from spherical symmetry of the planetary field. Usually this information comes from detailed study of the motion of the nearest satellites. With Uranus, the situation is greatly simplified by the presence of nine rings, one of which, the ε-ring, is clearly non-circular. From the motion of this ring, James L. Elliot, of the Massachusetts Institute of Technology, and his colleagues have been able to deduce extremely accurate values of these important gravitational parameters. For Neptune, the value, based on a study of the motion of its two satellites, is less precise.

In spite of the various uncertainties, several apparently satisfactory models of the interiors of Uranus and Neptune have been recently propounded by William Hubbard and James MacFarlane, of the University of Arizona, Morris Podolak,

The interiors of Uranus and Neptune, showing their marked similarity.

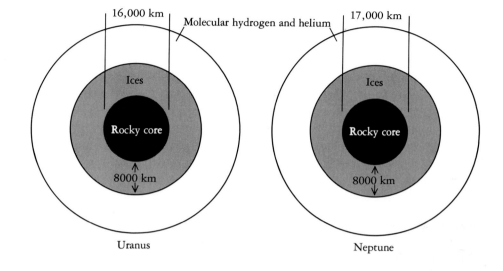

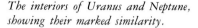

Orientation of Uranus as seen from the Earth in various years shows why a feature on its equator would have been visible in 1966 but not in 1985.

of Tel Aviv, and Ray Reynolds of NASA Ames. These models indicate that each planet has a dense rocky core—liquid or solid—of about fifteen Earth masses, which is about one-quarter of its total mass. About two-thirds of each planet—the mantle surrounding the core—is mostly water, partly liquid and partly ice, with some ammonia and methane, leaving only the outer ten percent for hydrogen and helium. The visible clouds are probably small crystals of ammonia and, perhaps, methane. The expected low content of hydrogen and helium shows the huge contrast between these two planets and Jupiter and Saturn. In the latter pair, hydrogen and helium constitute more than three-quarters of the mass of the planets; thus, the formation and evolution of the two pairs must have been quite different.

Much information about a planet interior can be obtained observing its magnetic field, if any, and its emission of heat. The situation with Uranus is somewhat paradoxical: It emits essentially no heat beyond what it gets from the Sun, but probably has a magnetic field. The existence of this field, suggested in 1976 by Larry W. Brown, of Goddard Space Flight Center, on the basis of weak and unconfirmed observation of electromagnetic radiation similar to that coming to us from Jupiter, was very recently made much more likely by observations of hydrogen light emitted from near the planet's poles. This light can be best accounted for by assuming that, around Uranus' poles, electrons trapped by the planet's magnetic field produce aurorae similar to but more powerful than those on the Earth. In contrast, Neptune emits three times the heat it gets from the Sun, but, so far as we know, has no magnetic field. It is of course, possible that one exists and is simply too weak to be detected from the Earth. This question should be settled by Voyager 2 in 1989.

All the current models of Uranus and Neptune suggest that they have rocky cores and that according to Michael Torbett and the author, the core of the former is liquid and metallic while that of the latter is solid. This conclusion agrees with the suspected presence of a magnetic field on Uranus, generated in its liquid core produced by heat from gravitational segregation of the heavier components—in a situation similar to that on Saturn. On Neptune there can be no similar mechanism because the core is mostly solid, but conceivably, a weak field might be produced in a thick layer of water, such as the one that surrounds the solid core, which at very high pressures would conduct electric currents like a metal.

The discovery of the rings of Uranus was a totally unexpected result of a fairly routine attempt to measure exactly the planet's diameter. Such studies are usually done by observing the motion of a planet across the sky and measuring the length of time during which the light coming from a particular star is totally obstructed by the planet. From the velocity of the planet, with respect to the background of the distant stars, and the duration of the occultation, one can deduce the planet's size. One can also get information about its atmosphere. James L. Elliot and his colleagues at the Massachusetts Institute of Technology while making such measurements of Uranus from an airplane, noted that a few minutes before the planet was to obscure a star, the star's light suddenly dimmed and, after a few seconds, reappeared. It did this five times before, finally, the planet itself occulted the star for several minutes. After the planet passed out of sight and the star became again visible, the five drops in the star's brightness occurred again, exactly in the opposite sequence. After eliminating all possible accidental instrumental sources of these phenomena, the observers concluded that Uranus had at least five rings, because a ring would occult the star twice, and a satellite, only once. Further studies confirmed this result and raised the total number of rings of Uranus to nine.

In spite of many attempts, a direct observation of the rings of Uranus in reflected visible sunlight similar to observations of Saturn's rings has proved to be essentially impossible, though they have been photographed in non-visible infrared radiation. This result agrees with the conclusion that these rings are narrow—from a few to 80 kilometers—and secondly, that they are dark, perhaps 40 to 50 times darker than the snow or ice that cover the rings of Saturn and make them easily observable. If the Uranian rings were directly observable, they would appear as a halo around the planet as we see it now, from above its pole. The ring nearest Uranus has a diameter one and one half that of the planet, while the ring farthest from it is about twice its diameter. Some of the rings are

Detail of a record of an occultation of light from a star by the ε ring of Uranus. The gap in the normal level of intensity of light from the star indicates the width of the ring.

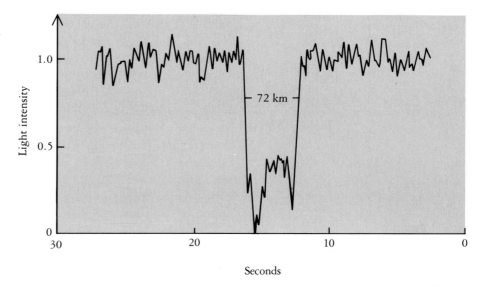

almost circular, others, elliptical, but all seem to lie close to the equatorial plane of the planet. Most likely they are remnants of a stray body, perhaps a comet or a stony asteroid, that came too close to the planet and broke up. Initially, it was thought that the location and width of the rings could be accounted for by resonances; a simple numerical relationship between the orbital periods of the ring particles and the periods of some of the satellites. These simple relationships do not seem to hold, however, and this result is rather surprising because the satellite Miranda, which has a diameter of 550 kilometers, is only 2½ times farther from the planet's center than the outermost ring and, thus, one would have expected a strong gravitational interaction with it. It is striking that Saturnian rings are many tens of thousands of kilometers wide, with narrow gaps between them, while those of Uranus are, at most, a few tens of kilometers wide, with gaps of a few hundred to a few thousand kilometers. Peter Goldreich and Scott Tremaine of the California Institute of Technology have proposed that, not unlike the F-ring of Saturn, each of the Uranian rings is held in place by shepherding satellites, which themselves may be the source of the ring material. Clearly there is still much to learn about the origin and structure of planetary rings.

Neptune seems to be a more normal planet than Uranus because its axis of rotation is inclined to the orbital plane a little more than that of the Earth, and so it may have seasons in the upper, thinner part of its atmosphere. Its distance from the Sun is so great that Neptune takes 165 years to complete its orbit and has not had time to go even once around the Sun since its discovery in 1846

culminated a pursuit that began in 1781 when Uranus was accidently discovered by astronomer Wilhelm Herschel in England. When French astronomer Alexis Bouvard tried to calculate the orbit of Uranus he found that it deviated from an elliptical orbit. After many unsuccessful attempts to explain this discrepancy had been made, French and English mathematicians Urbain Le Verrier and John Couch Adams suggested that it could be explained by the presence of another planet farther away, whose gravitational attraction perturbed the path of Uranus. On this basis, they predicted the location of Neptune, which was confirmed shortly by observations made in Berlin by Johann Galle. Actually, luck played a part, because, as we know today, both Le Verrier and Adams had used rather questionable assumptions in their calculations.

One of Neptune's satellites, Triton, is about 3500 kilometers in diameter, but we know essentially nothing more than that it revolves Neptune in a direction opposite the revolution of the planet. Another satellite, Nereid, is small—500 kilometers in diameter—and circles the parent planet in the normal direction but on a rather elongated orbit. In 1982, William B. Hubbard with his colleagues at the University of Arizona and Edward F. Guinan and his collaborators at Villanova University concluded that the planet may have a third, so far unnamed, satellite about 180 kilometers in diameter or a ring. Thus Neptune, the last major planet long thought to have no rings, may, indeed, have one. If the spacecraft Voyager 2, which is now on its way to Uranus, is still functioning seven years from now, it may be able to answer this question unequivocally.

PLUTO

Small and about forty times farther from the Sun than the Earth, Pluto is the least known and strangest of all planets. Its size is very difficult to establish and, at present, 4000 kilometers, or about the distance from Washington, D.C., to Los Angeles, is thought to be the best estimate. Measuring Pluto is like trying to measure a golf ball at a distance of 50 kilometers. While the orbits of the others lie within two or three degrees of one plane, Pluto's is inclined 17 degrees to it. This orbit is also the least circular of all planetary orbits; in fact, Pluto is sometimes nearer the Sun and sometimes farther from it than its neighbor Neptune—but the two planets never come close enough to collide. Since 1978 Pluto has been within the orbit of Neptune and will be at its closest to the Sun and to us in 1989. After the year 2000 and for the rest of its orbital period, which is 248 years, it will be again the most distant known planet. Pluto and Neptune are the only pair of planets for which such a periodic interchange of closest

Inclination of Pluto's orbit to the plane of the ecliptic (plane of the Earth's orbit) is about 17 degrees; all other planets lie within 2 or 3 degrees of it.

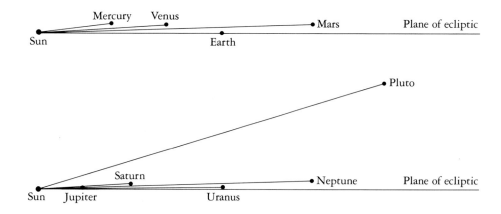

Eccentricity of Pluto's orbit permits it to intersect Neptune's. The dashed portions of orbits fall below the ecliptic.

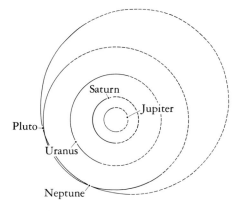

approach to the Sun does take place. It has been speculated that Pluto was previously a satellite of Neptune, and that the gravity of a passing body pulled it into a solar orbit. Whether this is indeed an acceptable description of the origin of Pluto has not yet been answered by detailed calculation.

The discovery of Pluto in 1930 by Clyde Tombaugh, at the Lowell Observatory, was preceded by an extensive search for a distant trans-neptunian planet that would explain the significant irregularities in the motion of Neptune and those slight irregularities of Uranus unexplained by the presence of Neptune. This procedure was similar to that which, described above, led to the discovery of Neptune itself a few decades earlier. Ironically, Tombaugh's success was due to his perserverance and not to the theoretical prediction of Pluto's path, which was incorrect. In addition, Pluto's mass—about $\frac{1}{300}$ that of the Earth—turned out to be at least one hundred times too small to account for the irregularities in the motion of Neptune and Uranus. We thus have now before us two puzzles: What is the origin of Pluto and its unusual orbit, and how can we account for the observed perturbations of Neptune and Uranus? So far no trace of a suitable trans-plutonian body has been found. The trouble is that the magnitude of the irregularities, if they are real, requires as their source a rather big planet, which should be easily observable with present-day telescopes; none, however, has been found. The irregularities could be due to a much bigger but darker planet much farther away or—as some courageous astronomers suggest—even a black hole. There is a very slight chance that the Pioneer 10 spacecraft, which is now on its way between Uranus and Neptune, and Pioneer 11 will cast some light on this tantalizing matter.

In 1978, James W. Christy and Robert S. Harrington of the U.S. Naval Observatory, noticing a small bump on the side of Pluto's image, concluded that

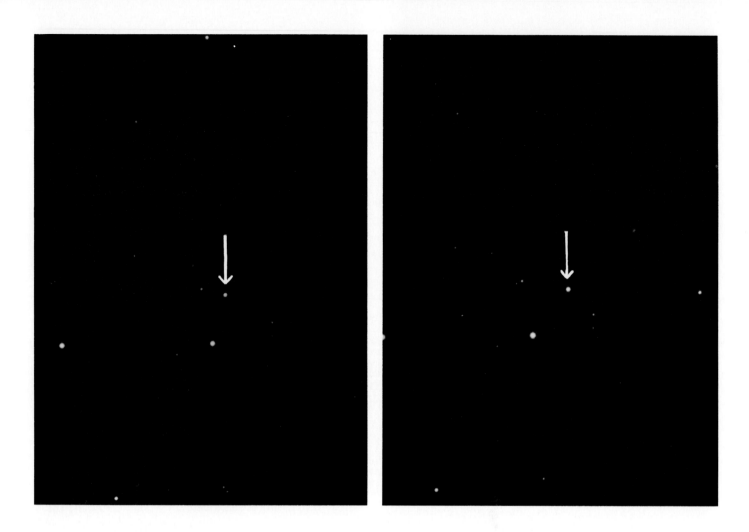

Two photographs, made one day apart, showing movement of Pluto (arrows) relative to stars, the phenomenon by which Pluto's existence was first determined.

it had a satellite, which they called Charon. (In classical mythology, Charon is the boatman who ferries the souls of the dead across the river Styx to Hades, the underworld ruled by Pluto.) The Christy-Harrington discovery was confirmed by subsequent observations in spite of the fact that, if Pluto is difficult to observe, Charon presents still more formidable difficulties for even the most sophisticated telescope and instruments. The best present estimates are that Pluto's mass is about one-fifth that of the Moon, and Charon's is one-tenth Pluto's. The period of revolution of Charon around Pluto is 6 days and 10 hours, which seems to be also the period of rotation of the planet so that with respect to Pluto's surface, Charon is stationary in the sky. Charon, about 20,000 kilometers from Pluto, is relatively big, about half the size of Pluto. The next biggest ratio of the size of a satellite to its planet is that of our Moon to the Earth, which is about one-quarter. Thus, Pluto and Charon look very much like a binary planet.

Optical studies show Pluto has an atmosphere that contains methane. According to Laurence Trafton of the University of Texas, some argon must also be

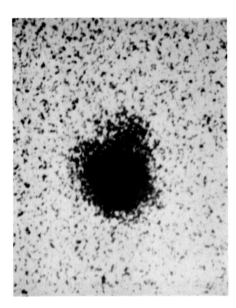

Pluto, the main mass at the center of the picture, and its satellite Charon, here appearing like a bulge (upper right) in Pluto.

present, preventing the methane from escaping the weak gravitational field of the small planet. The surface temperature of Pluto is near 42 K ($-231°$ C) which is much below the melting temperature of methane, so that the planet, or at least its surface, is probably a mixture of solid methane and other ices. This conclusion agrees with the average density of Pluto, which is perhaps half that of water and close to the density of solid methane, but much below the density of any other non-gaseous planet. Of course, the uncertainty about Pluto's size and mass make this conclusion very uncertain. Unless the calculated rate of escape of methane from Pluto is erroneous, the mass of this tiny body will gradually decrease, and eventually Pluto may disappear altogether. This entertaining possibility was suggested—tongue-in-cheek—by Alexander Dessler of Rice University and Christopher Russel of the University of California, Los Angeles.

Pluto is thus perplexing: It has an odd orbit, it is minute, and, unlike other solid planets, it seems to be made of frozen gases. For these reasons, some astronomers, such as Brian Marsden of Harvard University, lean toward omitting Pluto from the list of regular planets, thus acknowledging its mysterious origin and peculiar properties. It is tempting to put Pluto in the class of asteroids, some of which may have close satellites, or to call it an enormous captured comet, which is slowly evaporating, but neither of these possibilities really helps to solve the puzzle.

ASTEROIDS

Planets are spaced fairly regularly around the Sun but there is an obvious gap between Mars and Jupiter. All the various rules proposed for calculating and correlating the distances of the planets from the Sun suggest that there should be a planet at a distance about 2.8 times that between the Earth and the Sun. Instead of a planet we find there a wide belt of many bodies smaller than a few hundred kilometers in diameter. The first asteroid was discovered in January, 1801, by Sicilian priest and astronomer Giuseppe Piazzi, who initially thought he had found a comet; he named it Ceres, for the patron saint of Sicily. Ceres turned out to be the biggest asteroid—about 1000 kilometers in diameter. The discovery of Ceres had a special importance in the study of planets because it offered the first opportunity for German mathematician Karl Friedrich Gauss to apply his theory that the orbit of a body can be calculated on the basis of only a few observations. The accuracy of his calculation was so high that after Ceres had been hidden from view behind the Sun for some time, it was found again in the predicted position in the sky.

Distribution of asteroids relative to the Sun, and the location of resonances with Jupiter.

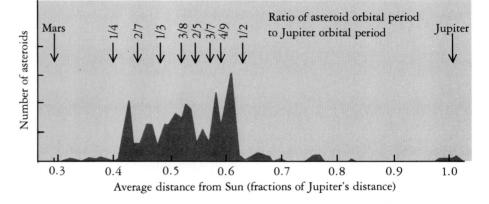

Altogether more than 2000 asteroids have been so far identified, new ones continue to be found and the total is probably in the tens of thousands. Since small bodies at this distance are difficult to observe, we are not sure if there is a lower limit to their size, but there is no reason why they should not be as small as a centimeter or so. It is thought that asteroids of a few ten-thousandths of a centimeter form, together with some cometary debris, a layer in the plane of the orbit of the planets. This layer reflects sunlight and is sometimes visible in the sky after sunset or before sunrise, usually as a rather faint luminosity, called zodiacal light, directed along the plane of the Earth's orbit. At a point opposite to where the Sun is one can see a similar brightness, called gegenschein. The total mass of all asteroids is, however, less than the mass of the Moon, and a large fraction of it is the large Ceres. Although the majority of asteroids circle the Sun in the belt between Mars and Jupiter, their orbits do not fill this space uniformly, and there are definite gaps in positions (or, rather, in periods), the so-called Kirkwood gaps, where there are no asteroids. The positions of these gaps agree very well with resonances with Jupiter—that is, with locations at which an asteroid's orbital time is a simple fraction of that of Jupiter. As a result, at these locations Jupiter and an asteroid align periodically, at which time Jupiter's gravity pulls the asteroid in the same direction so as to form a gap. Several mechanisms for the formation of these gaps have been proposed. Michael Torbett, of the University of Texas, and the author have shown that Jupiter's powerful gravitational influence, acting during the evolution of the solar nebula, is the reason why no major planet formed between it and Mars. Jupiter, because of its gigantic size, raised havoc in its vicinity by upsetting the systematic and slow processes that elsewhere led to the formation of the major planets. The motion of resonances, through the asteroid belt, generated by Jupiter's presence, changed the orbits and velocities of the particles to such an extent that, on collision, they broke up rather than accreted and were thrown from the belt. The

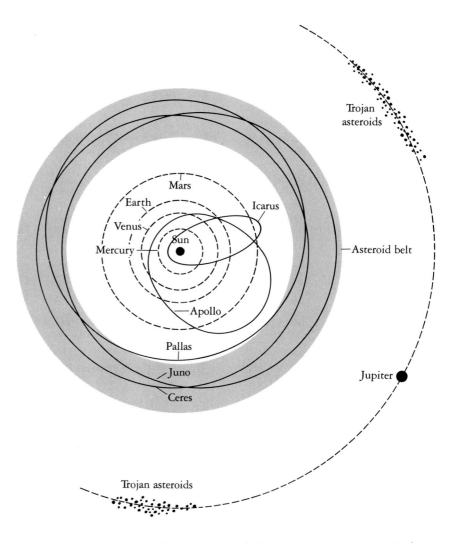

Most asteroids form a belt between Mars and Jupiter. Some penetrate much closer to the Sun, crossing the orbits of the terrestrial planets. The Trojan asteroids precede or trail Jupiter in its orbit by two-thirds of a right angle.

widths of the gaps calculated on the basis of this assumption agree with observation. Jupiter also forced some of the asteroids to circle the Sun along its own orbit. These, the so-called Trojan asteroids, either precede or trail the planet, and are located near points at which the various forces acting on them balance out. These positions were predicted theoretically by French mathematician Joseph Louis Lagrange at the end of the eighteenth century.

About 35 known asteroids cross the orbit of Mars. There are also about fifteen known asteroids—the so-called Apollo group—that cross the orbit of the Earth, but their actual number is probably several times greater. The chance that one of these asteroids will hit the Earth is small, but not entirely negligible. While the present orbits of the known asteroids do not suggest imminent collision, these orbits are being slightly altered all the time by the gravitational influences of the planets, and so there is no reason why some asteroids should not eventually make an impact on our Earth. It is estimated that the probability of such a collision with an asteroid bigger than half a kilometer is about one in 6 million years.

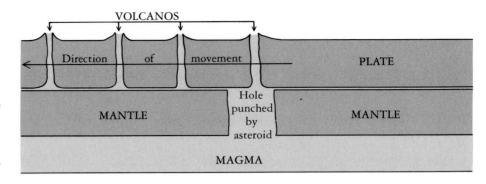

Schematic representation of possible role of an asteroid in creation of the underwater chain of volcanos from Kamchatka to Hawaii. The hole in the mantle, through which the magma rises, may have been produced by the impact of an asteroid.

Luis W. Alvarez, Walter Alvarez, and their colleagues of the University of California at Berkeley have suggested that there is evidence that some 65 million years ago an asteroid with an estimated 10-kilometer diameter hit the Earth, probably in an ocean. This event covered the Earth's surface with rare elements such as iridium and produced major temporary climatic changes. These changes may have led to the extinction of large animals, including the dinosaurs, which, fossil evidence indicates, disappeared from the Earth at that time, after 150 million years of existence. Anomalous amounts of these rare elements are indeed found in layers much less than an inch thick in many places all over the Earth at appropriate depths below the surface. According to Thomas Ahrens of the California Institute of Technology and John O'Keefe, huge amounts of water and rock were pulverized by the impact and carried high into the atmosphere, leading to a greenhouse effect, which produced a temporary significant increase of the average temperature on the Earth. Also, the high-altitude protective ozone layer was destroyed, permitting deadly ultraviolet radiation to reach the Earth. The atmospheric dust could also have decreased the amount of solar radiation reaching the Earth's surface for a few months, resulting in darkness and a drop in temperature. These effects may account for the fact that nearly half of all species of animals in oceans disappeared at that time. Whether these effects can be actually explained by the collision with an asteroid is not yet settled, but whatever the event, it was indeed a gigantic one.

Fred Whipple, of Harvard University, suggested that the formation of Iceland may be the result of an asteroidal impact 65 million years ago. At the University of Texas, Harlan J. Smith and the author drew attention to the underwater ridge of extinct volcanos stretching from Kamchatka to Hawaii and believed to be caused by magma (molten rock) rising through a hole in the mantle and periodically puncturing an overlying plate that is moving slowly in a northwesterly direction. The speed of the plate and the length of the ridges suggest that 60 to 70 million years ago the hole in the mantle may have been made by an asteroid. In 1937, the asteroid Hermes, which is 1.2 kilometers in diameter, came within 700,000 kilometers of the Earth, or less than twice the distance to the Moon. When the Earth-orbit-crossing asteroid Icarus, 1.4 kilometers in size, comes close to the Sun, it is nearer to it than Mercury, and, in

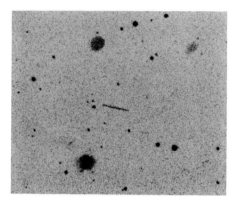

Trail of the asteroid Hathor, which came within 950,000 kilometers of the Earth in 1976.

1968, it came within 6 million kilometers of the Earth. A similar passage by another asteroid, but at a greater distance, occurred in 1972. Some asteroids have very elongated orbits, so that the most distant part of the orbit of Hidalgo, for instance, lies outside the orbit of Saturn; in fact our whole planetary system is continually crisscrossed by them, as well as by other objects. Calculations indicate that an impact on the Earth of an asteroid of 10-kilometer diameter could occur, on the average, every 100 million years.

The voyage of the spacecraft Pioneer 10 was the first beyond Mars, and there was much concern that, in crossing the belt of asteroids between Mars and Jupiter, it might be damaged by collision with one or more of the few thousand sighted asteroids and the many smaller ones we cannot see. Although Pioneer 10 and all the other subsequent spacecraft survived the passage through the asteroid belt, instruments on board did detect about 40 small but non-destructive impacts. There is no general agreement about the origin of the present size and distribution of asteroids. Undoubtedly, accretion and various effects of solar radiation tended to eliminate the smaller ones.

Asteroids are usually so small that estimates of their size involve risky assumptions. The best way to determine an asteroid's size is to observe it as it passes between us and a distant star and to measure the duration of the drop in the light reaching us from the star. Such a measurement was made in 1978 by Edward Bowell and his colleagues at the Lowell Observatory, who observed the medium-sized asteroid Herculina—diameter about 200 kilometers—in this fashion. To their surprise, the light of the star suddenly dipped a few minutes before the asteroid itself occulted the star. This result, which recalls the finding of the third satellite of Neptune and of the Uranian rings, suggested a satellite of about 25 kilometers in diameter. Whether Herculina, and perhaps other asteroids, really have small satellites is still not certain. If these satellites exist, they may have been acquired in the early stages of the evolution of the asteroid belt.

The surfaces of asteroids vary greatly. Some are of fairly dark rocky and carbon-containing materials, which reflect only a small amount of sunlight. Others reflect as much as 40 percent of the light falling on them. Some asteroids seem to contain iron and nickel at their surfaces, but these may be fragments of a larger asteroid that broke up, exposing a metallic center. These studies, together with the analysis of the orbits of asteroids, seem to indicate that the once popular theory—that the asteroids are remnants of a planet that broke up—is incorrect. On the other hand, a few groups of asteroids have almost identical parallel orbits and very similar surface chemistry. For instance, the Trojan asteroids seem to differ in their surface composition from all other asteroids. Quite possibly each of these "families" is a group of remnants of an asteroid that broke up long ago on collision with another asteroid or a passing external body. Phobos and Deimos, the two satellites of Mars, are very likely captured asteroids. Their density and surface chemistry are strikingly similar to those of most of the asteroids and quite different from those of Mars. Thus, the very good pictures of these two satellites obtained by Mariner spacecraft could quite possibly be what some asteroids look like at close range.

In 1977 Charles T. Kowal, of the Hale Observatories, found a minor body that, during its elliptical motion around the Sun about once every 50 years, passes close to Saturn and then recedes as far as Uranus. This body has been named Chiron, in Greek mythology the son of Cronus, or Saturn, and the grandson of Uranus. Is Chiron as asteroid or a comet—that is, is it made of meteorite-like materials or of ice and, where did it come from? To explain its brightness it would have to be less than 100 kilometers in size if made of ice, and 300–400 kilometers across if composed of dark material, like an asteroid. No one has observed an asteroid that came no nearer to the Sun than Saturn or that was as big as 100 kilometers, and there are no other asteroids between the orbits of Saturn and Uranus. Thus, Chiron is indeed a unique body, no matter what its composition. Its orbit is not stable so that it will probably leave its present location, and the proposal that it is an escaped satellite, either of Saturn or Uranus, is not impossible.

COMETS

Although casual observers may underestimate the significance of celestial phenomena that occur frequently and regularly, rare or unpredictable events arouse great curiosity, and the unsophisticated have tried to associate them with unexplained and sometimes feared occurrences. Thus, the regular motion of the starry skies, of the Sun, the Moon, and even of the easily visible planets create considerably less general interest than comets or meteors. Although some comets are seen in the skies quite regularly, their periods are very long and their previous appearances often forgotten. For instance, the soon-to-be-visible again Halley's Comet last seen in 1910, has made some 29 recorded appearances, approximately 76 years apart, since 239 B.C. A comet seen in 1057–58 by Chinese scholars may also have been Halley's Comet. Its appearance in 1456 prompted Pope Calixtus II to order the faithful to fend off the comet, "the anger of God," by a special prayer: "Lord save us from the Devil, the Turk and the Comet."

The regularity of the appearances of this comet escaped attention until British observer Edmund Halley discovered it in 1705 while comparing the orbits of comets seen in 1531, 1607, and 1682. One of its earlier apparitions, in 1066, is illustrated in the famous Bayeux tapestry depicting the Norman conquest of

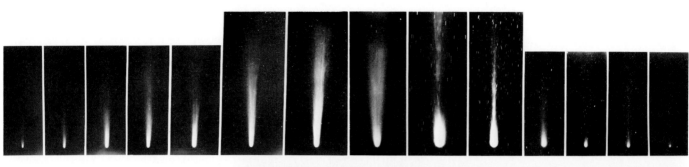

Sequence of photographs showing Halley's Comet as it approached and then receded from the Earth in 1910. At right is a detailed view taken at its closest approach. The coma has a 10- to 20-kilometer-diameter solid nucleus.

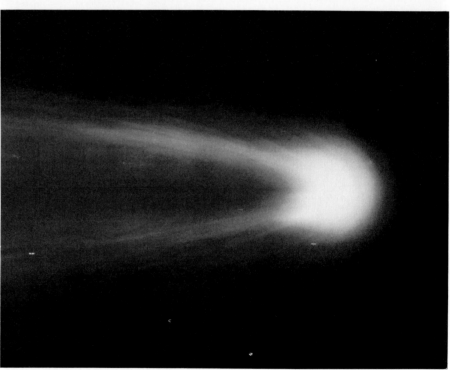

England. No doubt contemporary astrologers exploited this event and gained many followers. Its appearance in 1301 inspired Florentine painter Giotto to show it in a surprisingly realistic manner in his fresco The Adoration of the Magi. Naturally, there is general interest in catching sight of Halley's Comet early in its 1985–1986 approach to the Earth, particularly before it acquires its big shining coma and tail. On October 16, 1982, D.C. Jewitt, of the California Institute of Technology, and colleagues, using the 200-inch Mount Palomar

Halley's Comet, shown (top center) in the Bayeux tapestry, which records events that occurred in 1066. The legend preceding the image of the comet and referring to the figures at the left, reads, "They marvel at the star."

telescope and the most sophisticated electronic detectors, did indeed find a tiny speck that moved close to the predicted path. The speck was, however, so faint that in order to see it, it was necessary to block the light from nearby stars. At that time, the comet had passed the orbit of Uranus and was near to that of Saturn. Interestingly enough, the comet seemed to be about half a day late, which, in view of its perilous journey, is perhaps understandable. Its reflectivity indicates a diameter of a few kilometers.

What are comets and where do they come from? Earlier, in the description of the process that led to the formation of planets from the solar nebula, it was mentioned that the initial diameter of the nebula was much greater than that of the present planetary system, even including Pluto. It is possible that, as the nebula cooled in its outer boundary various gases solidified into icy aggregates of a number too small to permit the strength of the interaction needed to produce bodies comparable in size to the planets. Another possibility is that these icy aggregates were formed somewhere near Uranus or Neptune, which then ejected them to great distances from the Sun. The matter is, at present, not settled. In 1950, Jan Oort showed that these small bodies are indeed the source of comets. He made a careful study of the paths of comets, especially ones that had not before come close to the Sun and so had not been strongly influenced by it. All of these orbits are extremely elongated ellipses or parabolas, indicating that these new comets originated at distances 1000 to 100,000 times greater than the

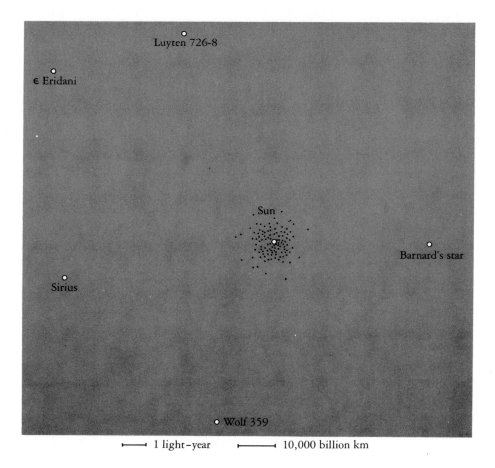

Luyten 726-8

ε Eridani

Sun

Barnard's star

Sirius

Wolf 359

├──────┤ 1 light-year   ├──────┤ 10,000 billion km

distance between our Sun and the Earth. Further calculations and estimates have
shown that in this region there are as many as 100 billion small bodies circling
the sun at a very slow 140 meters per second or less. These bodies make up the
Oort cloud, which forms a kind of spherical halo around the solar system, each
body in it taking a few million years to complete its orbit. Their total mass is
estimated about that of the Earth.

It has also been shown that about once in 100,000 years a star passes the Oort
cloud at a distance that may be ten times the cloud's distance from the Sun and
perturbs the otherwise fairly regular and more or less circular orbits of a few
comets in this enormous storage space. As a result, some of these comets will be
sent either toward the Sun or out of the solar system altogether. The initial orbit
of a comet that approaches the Sun for the first time is, of course, extremely
elongated, but later, because of interaction with the bigger planets, it may
become smaller and more nearly circular. Jupiter, being very massive, can so
perturb the orbit of a passing comet that there is little resemblance between the
original orbit and that after the encounter. For this reason, it was long errone-
ously believed that comets originated in the vicinity of Jupiter. Actually, since

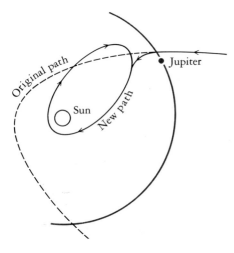

Typical alteration in the path of a comet as a result of perturbations of its orbit by a planet, in this case, Jupiter.

the paths of comets are always somewhat perturbed by planets, their orbits constantly change slightly, and so their returns are never exactly periodic.

The periods of comets vary greatly; among the better known, fewer than a hundred have periods shorter than 200 years and most of the other have much longer periods. Among the latter is Comet Kohoutek, which was seen for the first time in 1973–74 and, according to calculations, will not come back for 5 million years. This comet's orbit reaches so far that the velocity at the extremity of its orbit is only one kilometer per hour as compared to its velocity of nearly 500,000 kilometers per hour near the Sun.*

An icy body circling the Sun in the Oort cloud for millions or billions of years has a mean temperature of about 10 K ($-263°$ C). After it is perturbed and sent on its way toward the Sun, its temperature gradually increases. Finally, when the comet comes as close to the Sun as the orbits of Jupiter or Mars, the heat becomes so intense that the comet begins to evaporate, the evaporating matter forming a bright halo, or coma, around it and evenually also a visible tail. The signature of a comet is its tail, which is sometimes quite small but can extend from horizon to horizon. Actually, the coma is up to 100,000 kilometers in diameter and the tail may be 300 million kilometers long, or twice the distance from the Sun to the Earth. That the Oort cloud is occasionally perturbed by the passage of a star agrees quite well with the frequency of the appearance of new comets, taking into account both the brighter comets, which are easily visible to the unaided eye, and also the much more numerous smaller and more distant comets, which can be observed only with telescopes.

Until the solar light reflected by comets was analyzed with modern instruments, essentially nothing was known about their composition. We now know that a comet as it approaches the sun is composed mainly of ices—among them water, ammonia, and carbon dioxide in various proportions. Although we do not observe these cometary nuclei directly, we can identify the products of their break-up under the influence of the solar heat. The ices in a cometary nucleus are usually not solid, but porous, resembling snow, and include some dust, made of oxides, and other minerals. A comet, as Fred L. Whipple of Harvard University has said, is a "dirty snowball."

As a comet approaches the Sun and becomes progressively hotter, the most volatile components, such as carbon dioxide, start to evaporate first, followed by water-ice and perhaps other ices. The escape of these gases may be so fast that

---

* This difference is a consequence of the second of three laws of planetary motion formulated by Johannes Kepler: A line connecting the Sun and a planet sweeps through equal areas in equal times. Thus, a body in an elliptical orbit travels more slowly the greater its distance from the Sun; a planet four times as far from the Sun as the Earth, would have only half the orbital velocity and a period eight times longer.

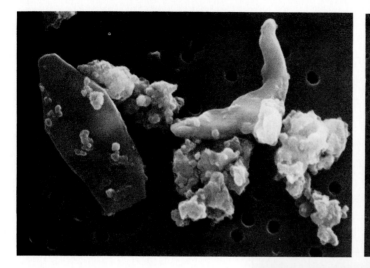

Particles, probably of cometary origin, collected in the Earth's stratosphere. Those at the top were collected at the time the Earth crossed the orbit of Halley's Comet. The dark particle at the left is .0005 centimeter long. The size of the very porous particle shown top right is indicated by the white line at the bottom, which is .0001 centimeter long.

The comet Mrkos displays a double tail in two views made five days apart.

they carry with them small solid flakes of dust and ice. At the same time, solar radiation and solar wind break up and charge some of the escaping molecules. These two processes, the flows of solid particles and of charged molecules, account for the sometimes double tails of comets: the fainter one being made of electrically charged molecules, which move along the lines of the solar magnetic field carried by the solar wind, and the brighter being made mostly of dust and ice particles pushed away from the Sun by its radiation. Both tails always point away from the Sun, so that the tail of a comet approaching the Sun is behind it, and that of one that has passed the Sun, precedes it, just as long hair streams with the wind no matter which direction a person is walking. In fact, the word "comet" derives from the Greek *kometes,* which means long-haired.

Each time a comet passes near the Sun it loses some of its volatile components through rapid evaporation, and so gradually becomes dimmer and dimmer and, eventually, invisible. Comets may also break up into many small pieces, which continue moving more or less along their original orbits. There is good evidence that the Earth passes periodically through such clouds of debris and that the small particles lead to the well-known meteor showers. An old comet that does not break up but simply ceases to emit gases and dust, may eventually become an asteroid and strike a planet or be captured by it as a satellite. As we have noted, the two satellites of Mars are probably captured asteroids, which, in turn, may be of cometary origin as are many satellites of the outer planets. There is fairly good evidence that the famous devastation of an area of Siberian forest of 2000–3000 square kilometers in the Tungusk region on June 30, 1908, was caused by a comet traveling at about 50 kilometers per second. The comet must have exploded at an altitude of a few kilometers, because no traces of a major stony or metallic meteorite or of a crater were found, which shows also that it was not a huge meteorite. Although there were few witnesses to the actual fireball, the explosion was so strong that it was heard some 1000 kilometers away, and, about 60 kilometers away, the shock knocked a man from his porch. There is evidence that the impacting body was a fragment of the Encke Comet.

The particles that reflect solar light in the dusty tails of comets eventually disperse in interplanetary space, and some may be attracted to planets. Within the last several years, Donald E. Brownlee, of the University of Washington, has been able to collect some of these tiny particles by capturing them on special plates carried into the atmosphere by high-altitude U-2 aircraft. These particles, a few ten-thousandths of a centimeter in size, no longer contain volatile substances but are composed mainly of various oxides and carbon compounds. Extremely fragile and porous, they are undoubtedly tiny fragments of much bigger particles that broke up on entering the atmosphere at altitudes of perhaps 150 kilometers or more.

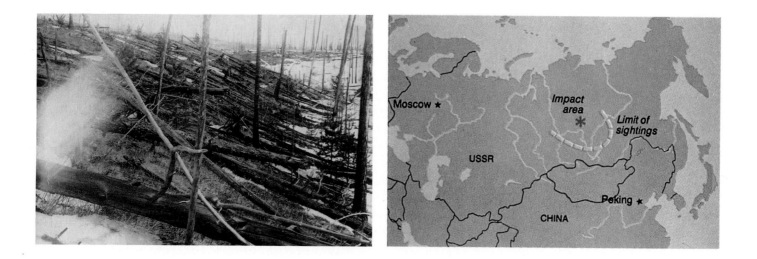

A small portion of the forest in the Tungusk region of Siberia that was leveled presumably by a cometary impact in 1908. The map indicates the impact area and the extent of sightings of the event. The photograph was made 20 years after the occurrence.

The European Space Agency and Japanese and Soviet scientists plan missions to observe Halley's comet from a close distance when it reappears in 1986. The results of such studies will throw light on the composition and nature of materials that condensed in the outermost reaches of the solar system billions of years ago.

That most comets are made primarily of water ice is the basis of a highly speculative proposal that has fascinating ramifications. It has been estimated by Max Wallis, of University College, Cardiff, that a reasonable content of radioactive elements in a 10-kilometer icy body with a 1-kilometer icy shell would lead to its maintaining a liquid water core for millions of years. Since liquid water is believed to be one of the essential factors for the maintenance of life as we know it, it is speculated that such a body could become a comet, disintegrate in the solar system, and perhaps be a carrier of some form of life into a planetary environment suitable for its development. Of course the proposal does not answer the basic question of how life entered the icy body in the first place.

Many comets come very close indeed to the Sun before continuing on their orbits. On August 31, 1979, however, American astronomers observed an astounding event—the first observed instance of a comet actually striking the Sun. The comet—known as the Howard-Koamen-Michels comet—had followed its elliptical orbit many times before without hitting the Sun. Undoubtedly, early in 1979 its usual path was perturbed by close passage near a big planet, probably Jupiter. This small change in its normal path was sufficient to bring the comet into the atmosphere of the Sun, which, in turn, offered enough drag to slow the comet to the point where it crashed into the Sun. Photographs made before,

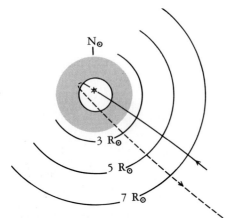

Course of the Howard-Koamer-Michels Comet, which crashed into the Sun in 1979. The dashed line indicates the course it would have taken had it survived the encounter, and the rayed circle is the point of impact on the Sun. Below, a sequence of photographs records the event. The numbers indicate 24-hour-clock time: "2032" was taken 57 minutes later than "1935."

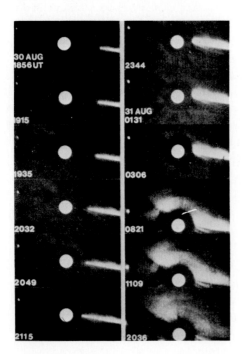

during, and after the collision leave no doubt of the event, and it is calculated that the comet disintegrated upon entering the Sun's atmosphere at an altitude of one solar radius—about 700,000 kilometers above the visible solar surface. The density of the Sun's atmosphere at this height is comparable to that of the Earth's at an altitude of 75–100 kilometers, which is where meteors approaching the Earth heat up and begin to be visible. No doubt the icy nucleus of the comet completely vaporized, and the non-icy debris was scattered into the solar corona, brightening it temporarily up to an altitude of 5 to 10 solar radii. Part of the comet's tail survived the collision and was gradually dissipated in space. It was altogether a spectacular event. Icy nuclei of new comets are usually several kilometers in diameter, so that they produce thick comas and tails when they approach the Sun nearer than about 500 million kilometers. Thus it was a surprise when, on April 25, 1983, the Infrared Astronomical Satellite reported that a new comet was already very close to the Earth and close to the Sun, but still not bright. Presence of good radar echoes suggests that only a fraction of a normal, few-kilometer-size nucleus is icy. The coma, 150,000 kilometers in diameter, was very thin, and there were only faint tails. Only once before was a comet seen at a comparably small distance from the Earth and, at that time, there were no instruments to make precise observations. It was possible, with this new comet, to obtain good data about its composition, structure, and motion within the coma that had never been seen. The orbit of the comet is almost an open one, so that after it passed the Sun on May 21, 1983, it may have left the solar system forever.

METEORS

Meteorites, found on the Earth, are the stony or metallic remains of meteors, or "shooting stars." Acceptance of the idea that these objects fall from the sky and are related to visible meteors was slow. The correlation was first suggested in 1794 by German physicist Ernst Chaldni, but it was not until a couple of decades later that the evidence became overwhelming. Thomas Jefferson, hearing a report on this matter by two professors from Yale, is reputed to have said, "It's easier to believe that Yankee professors would lie than that stones would fall from heaven." Often the fall of meteorites is accompanied by a loud noise or boom.

Usually, meteorites are remnants of much larger chunks of matter, called meteoroids, that, on entering the Earth's atmosphere, break up and become visible and so hot that their surfaces melt, burn, or even evaporate. The majority of meteors are produced by meteoroids that burn up within a few seconds of reaching the ground. The burning of small pieces broken off the meteoroid

Trail of a meteoroid (the thin line slanting upward from the right) in the sky above the Tetons in Wyoming.

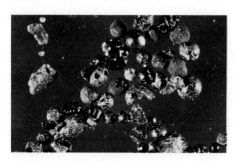

Meteoric dust particles from the bottom of the Pacific.

during its fiery entry into the atmosphere forms a trail, which may be quite prominent and occasionally visible even in daylight.

To realize that most of the Earth's surface is uninhabited water and that even bright meteors are usually easily visible only during dark moonless nights is to conclude that the actual number of meteorites reaching the Earth's surface must be enormous. In fact, estimates are that each day this rain of meteorites brings to the Earth about one million kilograms of matter, usually in the form of dust. The meteoritic dust particles are interesting because they often are slowed by the thin upper layers of the atmosphere to such an extent that they do not heat up much and thus do not change their internal chemistry and structure. This dust settles very slowly and is most easily found on the bottom of oceans.

The number of substantial meteorites reaching the Earth's surface is not small, even though many fall without being observed. In November, 1982, a rocky meteorite the size of a softball penetrated a house in Wethersfield, Connecticut, producing considerable damage. Eleven years ago, a similar meteorite crashed through another house in the same community. During the last 175 years, four such meteorites have been found in this one state.

Many huge meteorites have been found on the Earth, some lying on the surface and some buried in deep craters. One of the best-preserved and best-known is the Barringer Crater in Arizona, which is about 1.5 kilometers in diameter, 180 meters deep, and was made some 20,000 years ago—fairly recently by geological standards. Careful analysis of the crater and other geological evidence indicate that the meteorite disintegrated upon impact and covered the nearby countryside with more than 25 tons of mostly metallic fragments, rang-

An iron meteorite found on an ice floe in the Antarctic. Size may be judged from the centimeter scale at the bottom of the device at the left.

The Barringer Crater in Arizona, 1.5 kilometers in diameter and 180 meters deep, was formed little more than 20,000 years ago by a meteorite.

The Leonid meteor shower, which occurs late in November. Other similar showers are visible periodically.

ing from very small ones to some of a few hundred kilograms. This meteor left, however, no big chunks underground. At the time it entered the Earth's atmosphere, the original meteoroid was about 50 meters in diameter, weighed perhaps 500,000 tons, and was traveling at about 16 kilometers per second. On entry into the Earth's atmosphere, the meteor broke up; of the resulting fragments, the larger produced the crater in Arizona and a smaller one made a crater near Odessa in northern Texas. An even more famous "celestial stone" is the Black Stone enshrined in Kaaba, the Moslem sanctuary in Mecca, around 600 A.D. Many other craters have been found in remote parts of the world but few are well preserved.

Very likely, many meteoroids are products of asteroid collisions that occurred long ago; such collisions may still be taking place. A comparison of the optical properties of meteorite materials with those of asteroids supports this conclusion. Meteoroids that lead to meteor showers are undoubtedly remnants of comets and of their tails. For instance, the Taurid meteor showers are believed to be the debris of the comet Encke. In a few cases showers have been correlated with the disappearance of particular comets. Of course, some meteoroids could come from

140

Facing page. *Color-coded X-ray images of a remnant of a massive star that exploded in our Galaxy less than 350 years ago. Areas of highest intensities are shown in red and yellow.*

outside the solar system, but there is no good evidence of this. Whatever their origin, meteors are a proof that interplanetary space is full of all kinds of fragments of solid matter moving about, and we know from the detailed studies of the surfaces of atmosphere-free bodies, especially our Moon, that in the past the density of this interplanetary debris and, therefore, the rate of impacts were many hundreds of thousands times what they are today. What is particularly interesting is that the period of intense meteorite bombardment of our Moon occurred some 4 billion years ago and probably coincided with the cratering of Mercury and other terrestrial planets. The cratering of the satellites of Jupiter and Saturn may have had a different origin and timing; the matter is still unresolved. It seems, however, that the whole solar system was at some remote epoch subjected to intense bombardment.

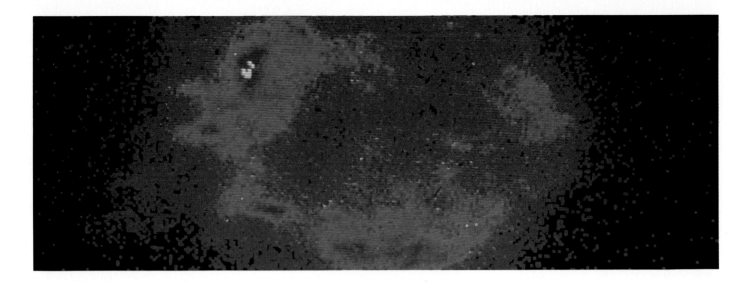

ESSAY III

THE FATE OF
THE SOLAR SYSTEM

The Sun will exist in more or less its present condition for another 5 billion years. At the end of that period, when a substantial part of the hydrogen near the Sun's center has been consumed, the fusion reactions will rapidly spread to the rest of the Sun, heating and expanding it enormously until it becomes a red giant. This giant star will envelop and evaporate all the terrestrial planets, including the Earth, thus erasing all traces of life, culture, and civilization, including the records of our knowledge of the universe. There are many such red giants in our galaxy.

Eventually, this red-giant stage of the life of the Sun will be over, and, after a few more evolutionary spasms, each connected with the onset of another kind of nuclear reaction, the Sun will begin to shrink rapidly and approach its end. During this shrinkage, the Sun's central temperature will rise for the last time. This will happen so fast that an enormous explosion will take place, blowing off into space most of the outer layers of the star

that was once our Sun. Eventually, the ejected gas and dust will cool and merge with interstellar clouds of dust and gas from which new stars will form. Like all other stars, our Sun literally rose from dust and gas and will end largely as dust and gas, closing the circle. Whatever is left of the central part of its red-giant stage will become a small, very hot, very dense star—a white dwarf. After a very, very long time a white dwarf, which has no source of nuclear energy, will gradually cool, glow like an ember, then finally become part of the dark matter in the galaxy. This will be the true death of our Sun.

During its life as a star, during the 10 billion years that it will have been the center of the solar system, after the red-giant and white-dwarf stages, and as a dark and cold remnant, our Sun will drift across its arm of our galaxy until it ends in the dark region between the arms.

The stellar life time of millions and billions of years prevents us from witnessing the whole life of any particular star.

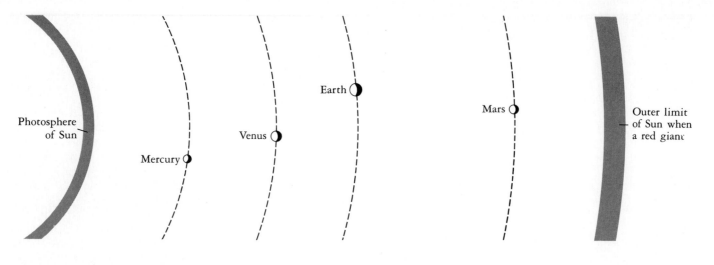

Photosphere
of Sun

Mercury

Venus

Earth

Mars

Outer limit
of Sun when
a red giant

*The Sun in its red-giant stage will expand
until it extends beyond the present orbit of
Mars.*

*The ring nebula in the constellation Lyra.
This nebula is the expanding matter blown
from a shrinking star, the core of which is
visible at the center of the ring.*

How then do we know that the outline presented here and in Essay I of the birth, life, and death of stars—and especially of our Sun—is true? Our belief rests on three important factors: All observations of cosmic phenomena suggest that the laws of physics, known from theoretical studies and from laboratory research, are truly universal. Using those laws, we are able to construct models of star interiors that explain their observed behavior. Furthermore, an enormous body of observational data concerning literally millions of stars permits their detailed classification according to size, composition, age, and energy output, which makes possible correlation of these variables with the models of stellar interiors, and their evolution, by means of which we can predict in some detail the life of a typical star. Finally, and most important, astronomers observe many, many stars illustrating each of the predicted periods in the life of a star: birth, maturity, onset of old age, spasms, explosions, and gradual disappearance from sight. Thus the process used to study the stars is, in a sense, similar to what would be the situation were we to observe all animal species for just a few seconds and then propose, on the basis of careful classification of data, and with some knowledge of biology, a likely chain of events, from birth to death, for each species. Of course stars, without the irrational and whimsical tendencies of animals, are incomparably easier to treat in a quantitative manner. As with all theories based on statistical inference, our theories of stellar evolution may not apply in detail to each particular instance, but those theories provide an exceedingly useful explanation of the norm and form the basis of astrophysics.

The only way I know of dealing with human conceit is to remind ourselves that man is a brief episode in the life of a small planet . . . and that. . . .other parts of the cosmos may contain beings as superior to ourselves as we are to jellyfish.
BERTRAND RUSSEL

AFTERWORD   LIFE IN THE SOLAR SYSTEM

Why is there life on the Earth when, as far as we know, there is none on the other planets? The question is part of a much broader issue: Does life exist anywhere else in the universe and, in particular, in our galaxy? We have begun to be able to give tentative answers to some of these questions but unambiguous statements are difficult because few facts are available and because the issues are sensitive, touching on the need of some to believe that we on the Earth are unique, and the rest a desolate immensity.

One can easily make a reasonable, semiquantitative argument that the number of stars in our galaxy, the Milky Way, is so enormous—about 100 billion—that some may have acquired planetary systems of their own. Of course, the same is true of the millions of other galaxies. If these stars were not too hot or too cold or too short-lived, the planets may have offered suitable conditions for the establishment of life, and a tiny fraction of them may have had conditions under which life, and intelligent life in particular, life could develop.

Observational search for other planetary systems is very difficult, but modern instruments begin to make it feasible. The crux of the problem is to ascertain whether the motion of a star and its velocity are smooth and regular or are perturbed, perhaps by planets. Critical are the size of the perturbations that could be expected and their observability at the enormous distances involved. If, for example, one were to observe our Sun from another star, the perturbations of the Sun's path due to the presence of the planets would be smaller than the solar diameter. In order to see these perturbations from the nearest star, Alpha Centauri, one would need a telescope capable of sighting from the Earth an object smaller than 10 meters on our Moon. This is not yet possible. A study of the variation of velocities of stars is much more promising, and about one-tenth of the stars similar to our Sun so investigated seem to show such irregularities. Of course, these variations may be caused not by planets, but by other invisible companions, such as neutron stars or dark dwarf stars (Essay II). In this connection, T. A. Heppenheimer, from the Center for Space Studies, has shown that it is very unlikely that Earth-like planets could be formed from the primitive nebulae surrounding stars that, as it is often the case, have sizeable companion stars. Two stars perturb and complicate orbits of particles around them, so that these particles have little chance to come together slowly and form larger clusters. This argument resembles that used to explain why there is an asteroidal belt rather than another planet between Mars and Jupiter.

Various equations that purport to evaluate the probability that life, and especially intelligent life, exists on other planetary systems do little more than stress our deep uncertainty about the origin of life and evolution. Depending upon personal preferences, scientists assume values of the various quantities and probabilities in these equations that may be a hundred, a thousand, if not a million times different from those assumed by others.

If there *are* other civilizations, some perhaps even more developed than ours, then why have they not tried to communicate with us, or if they have, why do we know nothing about their efforts? Is their search, like ours, beset by profound ignorance of what to search for? Some have speculated also that a civilization, say a billion years older than ours and located not more than 20 light-years away, might have detected the quite energetic radiation emitted by our TV stations since they became numerous and powerful. If this is so, then the question is really a different one: Why should these beings spend more time and effort trying to communicate with us than we do in trying to communicate with some less advanced species, say lobsters or fruit flies? A more basic question remains: What is the chance that another civilization reached its peak of technological development and has already ceased to exist, either through self-destruction or through a cataclysmic cosmic event? Maybe all civilizations, when they attain the capability and the will to search for other civilizations, also reach the stage when they are not capable of preventing their own extinction.

LIFE AS WE KNOW IT

Although we think we know much about life, we have no generally applicable definition of it. It is fairly easy to summarize the attributes of life: organization,

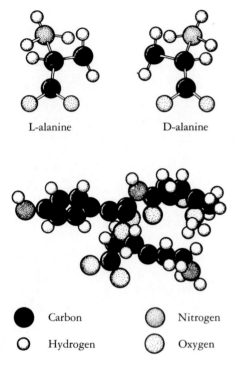

L-alanine   D-alanine

● Carbon   ◉ Nitrogen

○ Hydrogen   ◉ Oxygen

Above, a schematic representation of alanine, a typical amino acid. Amino acids may come in two forms, a left-hand (L) and right-hand (D), mirror images of each other. Below is a representation of three amino acids linked to form a segment of a typical protein.

ability to produce energy by using the environment, ability to grow, sensitivity to stimuli, adaptability that assures evolution, and, finally, ability to reproduce. All these attributes, including birth and death, have to have a mechanism based on molecular processes, especially on organic molecules, which contain carbon. This is so because, of all the known elements, carbon enters into, or is a backbone of, the greatest number of compounds. In fact, some 97 percent of the 6 million or so known compounds are organic, though not all of these occur in biological systems.

All living organisms are made of two kinds of basic molecular building blocks: twenty amino acids and five nucleotides. These twenty-five organic molecules, containing up to 30 atoms of carbon, nitrogen, oxygen and hydrogen are the same in *all* living matter—whether a tree, a fish, people, or bacteria. It is only the manner of arrangement of the amino acids—in complicated, lengthy units called proteins—that distinguishes one kind of living matter from another. Similarly, nucleotides link into long chains called nucleic acids, which differ among themselves in the sequence of the nucleotides. How are these arrangements produced and reproduced? The key is deoxyribonucleic acid (DNA), which is composed of linked nucleotides. In each living cell, whether a cell of brain tissue, of skin, or of a plant's leaf, there are DNA molecules, each consisting of two intertwined chains or strands of nucleotides—the famous double helix discovered by Francis Crick and James D. Watson. These chains, each containing a large number of atoms—sometimes as many as ten billion—are rather weakly bonded to each other and are supported by a kind of scaffolding of other molecules. It is the sequence of the nucleotides in each DNA chain that determines and preserves the genetic information. This sequence rather than the individual nucleotides determines the function of a cell—whether it is in a mouse or a tentacle of an octopus or the petal of a flower. When cells multiply during growth of an organism, or in order to replace dead cells, their DNAs too have to multiply, or replicate. In this process the weak bonds between the two strands of a DNA molecule break, "unzipping" it. Each of the two strands then uses free nucleotides available within the cell to synthesize a complementary strand, thus forming two complete DNA molecules, each having the same genetic information.

In the protein-forming process a special molecule attaches itself to a DNA molecule and opens it locally—that is, it separates the two strands for a small distance. Certain modified nucleotides, slightly different from those making up DNA, "read off" that part of the DNA, forming a sequence. the two strands then rejoin and the local separation moves along to the next section on the DNA chain where, again, the DNA is read off and another step added to the sequence. The process continues until a complete new molecule—ribonucleic acid

At right, a model of a DNA coil having 171 base pairs—about 7,000 atoms—in a computer-drawn image. Carbon is shown in dark blue; nitrogen, light blue; oxygen, red; and phosphorous, yellow. Hydrogen atoms are not shown. Below, an electron photomicrograph of a portion of an actual DNA molecule. The looped tangle, left center, is about .0001 centimeter wide.

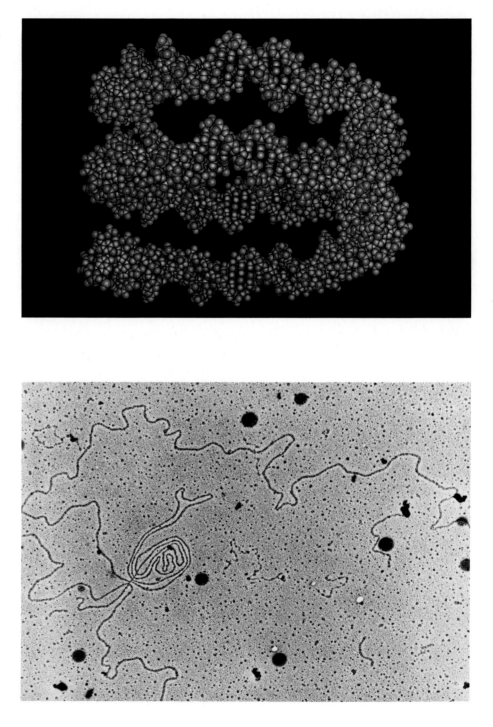

(RNA)—is formed that carries the same genetic information as the initial DNA, albeit encoded in a different form. With the help of other molecules, so-called RNA transfer molecules, various amino acids present in the cell's fluid attach themselves to the RNA, which thus becomes a template for the formation of a chain of amino acids—a protein. Repetition of this process leads to the formation of a large number of proteins, all identical, each having the imprint of the same genetic code. The large number of nucleic acids in a DNA molecule and of amino acids in a protein permits a mind-boggling variety of possible sequences or encodings of all kinds of characteristics of an organism. Such properties as eye color, food needed, size and number of limbs, or the ability to fly are all encoded, as sequences of the appropriate building blocks, in the initial DNA, in the RNA, and in the proteins formed. In this way, growth, inheritance of characteristics, function, and reproduction of life are assured. The details of these processes are quite involved and include factors that cause their eventual cessation: death of the cell or of the whole organism.

In our wildest speculation about life elsewhere we usually only rearrange or alter slightly what we already know. Imaginary beings in science fiction always resemble, in one way or another, human, animal or vegetable forms. One usually thinks of life in an environment in which there is water, free oxygen, and carbon-containing organic molecules, and where all the energy for growth ultimately derives from the sunlight. Actually, there are other possibilities, such as the recently observed bacteria in the deep (2500 meters), dark waters of the Galapagos Rift, which obtain energy from geothermal sources. We also know that there exist anaeobic bacteria, for which oxygen is not only not a need, but a poison.

It is quite certain that the early Earth, like the present Venus or Mars, had essentially no oxygen in its atmosphere and that this gas was produced on our planet by organisms that decomposed carbon dioxide into carbon and oxygen. Thus, anaerobic bacteria, which produce and reject oxygen, were probably, some 4 billion years ago, the first forms of life on the Earth. Eventually, oxygen-consuming organisms established themselves as the main form of life.

A life form that is to maintain its existence must embody a mechanism that offers the possibility of minor changes, or mutations. These can be the basis for adaptation to changing environment, leading to advances and evolution of the organism. Thus it is essential that DNA molecules undergo random mutations or accidental changes in the sequence of their nucleotides if the screening of the organisms by natural selection, and consequent evolution, is to take place.

It is noteworthy that carbon atoms lend themselves to the formation of a greater variety of intricate molecules than do atoms of any other element, such as silicon. We know also that certain elements important for biological carbon

Lightning and volcanic gases, such as may have combined to produce the first amino acids and the beginning of life on the Earth, fill the sky during the emergence of a new island in the sea off the coast of Iceland. The small red spots are photographic artifacts.

compounds are as likely to be encountered elsewhere in the universe as in the solar system. Thus, the chance of the existence of non-carbon-based life is very slim indeed, although life based on a somewhat different carbon chemistry is conceivable.

In 1971, Nobelist Charles Townes of the University of California and collaborators found radio-spectroscopic evidence of organic molecules in interstellar space. By now, the list of such molecules is long and includes many known to play an essential role in life chemistry. These molecules occur in dense galactic dust clouds, which protect them from the destructive stellar ultraviolet radiation. Other molecules, including those of amino acids and nucleotides, have been found imbedded in meteorites, which are undoubtedly of extraterrestrial origin.

In 1952, Nobel chemist Harold Urey and Stanley Miller, of the University of California, in experiments repeated by others in various forms, simulated prebiotic conditions that, presumably, existed in the early solar system, when organic molecules were already present but life as such had not yet appeared. In the experiments electrical discharges, that is, powerful sparks, were produced in a container holding a mixture of gases—such as ammonia, methane, water vapor, and hydrogen—common on the early Earth. The result was the appearance of certain kinds of amino acids and nucleotides. It seems thus that giant

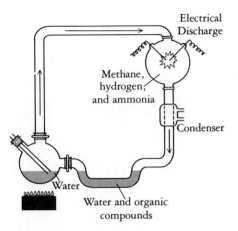

Electrical Discharge

Methane, hydrogen, and ammonia

Condenser

Water

Water and organic compounds

Schematic diagram of the apparatus of the Urey-Miller experiment, in which organic and even biological compounds were produced.

sparks—lightning—which have been observed in the atmospheres of various planets, could lead to the formation of some of the basic building blocks of life on planets with atmospheres containing these gases. Of course, these experiments indicate only the possibility of these processes and are not proof that life exists on these planets.

Are we justified in concluding from all these facts that life, however we define it, does exist somewhere besides the Earth? Some scientists answer with an enthusiastic affirmative; others are still uncertain: How did the various organic molecules, or biological building blocks, whose presence in interstellar space is now understandable, come together at the right time and in the right manner to form DNA, RNA, and eventually all the complicated proteins synonymous with the existence of life? A satisfactory answer does not seem to be available, although much progress has been made. The situation is rather like having a bag full of blocks, each inscribed with one syllable of the English language, and having them arrange themselves spontaneously to form a long sentence. Presumably, the original "sentences" in the early chains of amino acids and nucleotides were much shorter than the present ones and conveyed very little and probably ambiguous genetic information.

At present one of the favored explanations of how the complex, long spiral molecules were formed is that some amino acids and nucleotides in pools of water deposited a thin layer on a clay surface. In the course of time, those building blocks that were attached to the clay surface gradually aligned themselves to form simple chains that led to proteins and other biological structures. Experiments have shown that this can indeed happen and that the process is especially efficient if the temperature and the amount of water are changed in a periodic manner resembling that of the daily variation on the Earth. There is hope that this type of research will eventually clarify all the molecular mechanisms that are responsible for the origin of life.

Nobel biophysicist Manfred Eigen, of the University of Göttingen, and his colleagues have studied the likelihood of the formation of real genetic codes in DNA molecules rather than more-or-less random chains, and the statistical and thermodynamic laws of self-organization governing the natural selection of prebiotic molecules. They have shown that if "correct" genetic molecules were more stable and more likely than others to duplicate themselves, then long and meaningful "sentences" could spontaneously form through interaction of primitive proteins and nucleic acids. Thus, development of the function and of the structure of genes would have occurred simultaneously and there is no "chicken-versus-egg" problem. With respect to these processes John Hopfield, of the California Institute of Technology, has shown that there is a "proofreading" mechanism that minimizes the formation and duplication of "wrong" molecules.

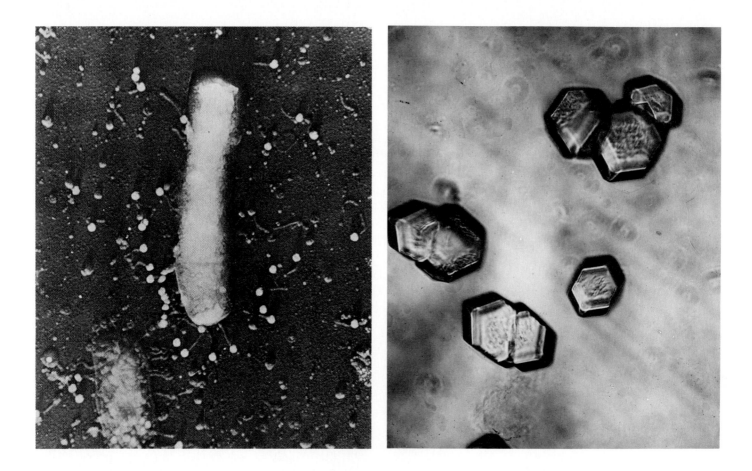

At left, viruses attack a bacterium. Fragments at lower left are remains of a bacterium attacked one hour earlier. At right are crystals of Coxsackie A 10 virus.

In sum, although many pieces of the puzzle of the origin of life are still missing, there are very promising lines of research that suggest how non-living chemical molecules can, under suitable conditions, become living proteins and organisms. Viruses, in fact, lie on the borderline of living and non-living matter, and seem at different times to show behavior of one or the other. For instance, the tobacco mosaic virus, which contains DNA molecules but no free amino acids or nucleotides, can exist for a long time in an entirely inert crystalline state. Dissolved and in contact with bacteria or other cells, it can introduce its DNA into them and destroy them. This destruction occurs when the DNA from the virus uses the amino acids and nucleotides of the invaded cell to produce proteins that duplicate the virus rather than the bacterium.

As far as the origin of life on the Earth is concerned, we have various kinds of evidence that suggest definite time limits. For instance, recent studies of certain rocks from Pilbara in Western Australia indicate that these rocks were formed some 3.4 to 3.5 billion years ago by algae and possibly other simple organisms that could trap rock-forming compounds. At that time the Earth was just over one billion years old. It is estimated that the Earth could not have been hospitable to any kind of life earlier than, say, 400 million years after its formation because of unsettled temperature, pressure, and surface chemistry. This conclu-

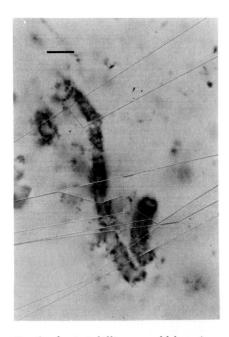

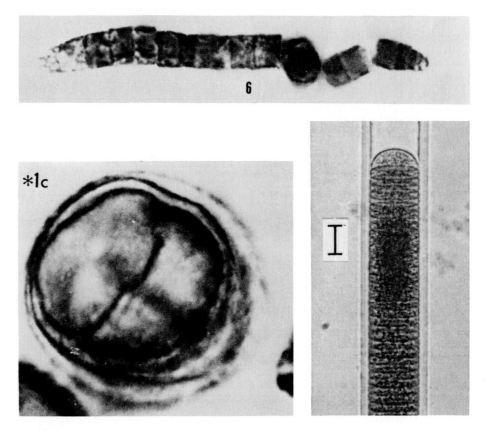

Fossils of a 3.5 billion-year-old bacteria-like organism found in Australia (above) and of multicelled organisms 850 million years old (right and top right) and 650 million years old (far right), all found in Africa.

sion indicates a period of roughly half a billion years for the start of the simpler forms of life on the Earth. The earliest organisms consisted of one or at most a few cells and probably did not need oxygen, deriving their needs from what is usually called the primordial soup—the mix of chemicals presumed to have existed in the first seas. In somewhat younger rocks—2.7 to 2.8 billion years old—whole organized communities or colonies of primitive algae were discovered that were more advanced in development and needed oxygen and thus light. Evidence of organisms showing sexual differentiation has been found in rocks dating to some 900 million years ago.

It has been suggested that life on the Earth started when spores or other organisms imbedded in interstellar or interplanetary dust and meteorites reached the Earth. Actually, several well known meteorites such as the Murray, Murchison, and Orgueil contain organic compounds that differ somewhat from those found on the Earth. Specifically those that have a screw-type structure differ with respect to their right- or left-handedness. While it is not certain where and how these compounds were formed, they are not strong evidence for the existence of life elsewhere. This suggestion, however, simply shifts the question of the origin of life from the Earth to space. Some 10 years ago Nobel biologist Francis Crick, of Cambridge University, and Leslie Orgel, of the Salk Institute in La Jolla, went even further and proposed that these "seeds" were sent to the Earth by intelligent beings from other planets. This idea seems even more esoteric because it presupposes that *intelligent* life, rather than just life, started somewhere else earlier. No evidence either for or against this possibility is available, and some scientists doubt whether we will ever have an answer. The strongest argument against it is the extremely small probability that a DNA molecule could survive long exposure to interplanetary and interstellar radiation unless it was enclosed in a substantial solid object.

What is special about the Earth that made it a place where life developed? Michael Hart of Trinity University has arrived at an important conclusion concerning this question by studying the evolution of the terrestrial atmosphere. Taking into account such factors as the probable changes in solar heat and in cloud coverage, he has shown that prior to 2 billion years ago the surface temperature of the Earth was some 30° C higher than now, because of a mild greenhouse effect. Had the Earth been nearer to the Sun during this period by only 5 percent, the greenhouse effect would have been much stronger and the surface temperature would have risen far beyond that at which life is possible. By about 2 billion years ago, enough oxygen had been produced and enough carbon dioxide locked up by vegetation and in the oceans that the mild greenhouse effect ceased; since then the Earth's temperature has been favorable for the evolution of life. Had the Earth at that time been only one percent farther from the Sun than it is now, all of its water would have been sequestered in glaciers, again eliminating the chance for life as we know it. Thus, life prospers on the Earth as the result of a delicate balance of orbital and chemical circumstances. One instance of *how* delicate is the fact that, were terrestrial vegetation suddenly to disappear, the Earth would lose nearly all its free oxygen and the atmosphere would become rich in carbon dioxide, as it is on Venus and Mars. The range of orbital radii that assures such an optimal climate on a planet is called a continuously habitable zone. For some stars, like our Sun, the width of such zones is reasonable; for others no such zones are possible. These limitations suggest that for life to exist, a planet has to be located in a certain temperature zone for a long enough time. Some scientists question Hart's assumptions and claim that the limits he proposes are too narrow. Others, such as James Lovelock and Lynn Margulis, of Boston University, argue that life itself, and especially plants, contribute to the maintenance of suitable climatic conditions on the Earth.

Variations in the width of the zone in which intelligent life could develop depending on the mass of the star. Zones shown are for stars whose masses are expressed as fractions of the mass of the Sun.

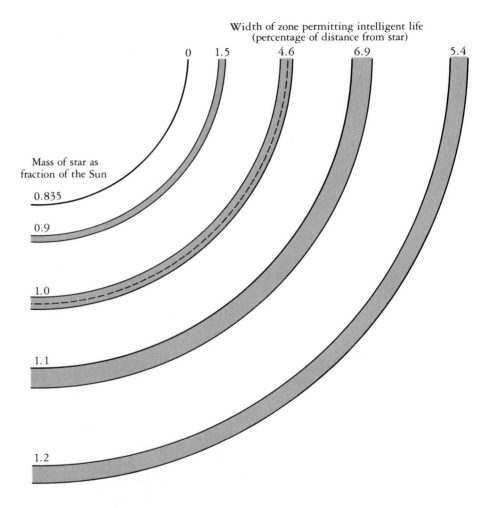

Width of zone permitting intelligent life (percentage of distance from star)

Mass of star as fraction of the Sun

0.835

0.9

1.0

1.1

1.2

The requirements for developing intelligent life are much more stringent than those for primitive life. It seems that the higher animals have a certain degree of intelligence, however one may define it, and this intelligence appears to be related to brain capacity. An intelligent animal avoids dangers and prolongs its life, leading to rapid natural selection. At what stage this process produced mankind as we know it is not known, although fossils dating to 1.76 million years ago found in Tanzania indicate the existence then of a bipedal hominid able to fabricate stone tools and able to climb trees for safety and for food.

The probability of finding life elsewhere in the solar system has been discussed by many scientists, especially in connection with the Viking missions to Mars, which included several experiments aimed at determining the existence of life on that planet. Many of us have had a rather romanticized view of the possibility of finding life on the surface of Mars, in the atmosphere of Jupiter, or on Saturn's satellite Titan; however, most of the arguments supporting this view do not survive careful scrutiny. Probably the most sober assessment has been given by Norman Horowitz, of the California Institute of Technology, who pointed out that the upper temperature limit for stability of amino acids is 475 K (200° C) to 575 K (300° C), that proteins break up at pressure higher than 4000 atmospheres, and that a planet supporting life has to have a gravitationally stable atmosphere to permit highly volatile decomposition products of dead organisms to be retained and recycled. The lower temperature limit is determined by the freezing point of the presumably liquid medium that various organic molecules must have in order to form and move. Interestingly enough, Soviet chemical physicist Vitali Goldanski has shown that, even at very low temperatures, close to absolute zero, where all thermal motion ceases, there is a measurable, though extremely slow, rate of protein formation by a mechanism called tunneling. It is not certain, however, whether this effect alters in any way the conditions described above for starting and maintaining life, which require continuous supply of energy either from direct sunlight or from other thermal sources.

Taking these various considerations into account, it appears that life cannot exist on Mercury because the planet has essentially no atmosphere and, in the past, its surface temperature was much too high. Venus has a thick atmosphere that could perhaps sustain some life, although it is very dry and, again, the planet's present surface temperature is much too high. Whether in the past, before the greenhouse effect developed, there was life on the surface of Venus no one can say. Even sampling of this surface, if feasible, may not give an answer, because the expected high erosion rate and volcanic activity are likely to have erased all traces of any life that may have been. On the other hand, it has been pointed out that, at altitudes of 30 to 70 kilometers, the temperature of the atmosphere of Venus is favorable to life, and some form of it could exist on, or in, cloud droplets, though they are made of concentrated sulfuric acid or dust in this region. The main problems with this suggestion are the persistent powerful winds on Venus and prevalent vertical mixing or convection, and ensuing evaporation and recondensation of the cloud droplets. To survive, an organism would have to have the ability to float within a certain range of altitudes and to resist the winds. A suggestion with a strong science fiction flavor was that such organisms might have the form of small balloons that adjust their size and buoyancy to stay at an altitude of suitable temperature.

Mars was and still is considered by many the most likely planet on which life could be found. The main reasons, apart from its general similarity to the Earth, are that it has an atmosphere that, though thin, is not negligible, and that the temperature is favorable. The possibility of vegetation on Mars was, for decades, the mainstay of these arguments. There is, however, practically no water in the planet's atmosphere and little molecular nitrogen, because these molecules are broken up by solar radiation into free atoms, which are too light to be held by the planet's gravity. These negative aspects have been buttressed by the biological experiments of the Viking mission, for no trace of present life in any form was found, and all the experimental results could be interpreted in terms of inorganic reactions. This was, of course, very disappointing from scientific, planetological, and emotional points of view.

There remains the tantalizing possibility that long ago the climate of Mars was warmer and liquid water abundant—so that life may have existed. When the present ice age on Mars passes, some form of life may reappear. At present we are unable to assess the probability of past or future life on this planet, but there is hope that before too long we will be able to bring geological samples from Mars—especially from the layered regions near the poles—and examine them for traces of early life.

The giant planets Jupiter and Saturn have no solid or liquid surfaces, and so life would have to exist in their atmospheres, in the clouds or on small dust grains, or, as proposed for Venus, "balloons." Both Jupiter and Saturn have atmospheres with regions near room temperature and both are rich in the gases in which electric discharges produced, in the laboratory, amino acids and nucleotides. When the Voyager missions observed huge electric storms on both planets, speculation arose that there might be life on them. The chief difficulty, however, is how the various organic molecules can assemble to produce DNA and proteins while floating in an atmosphere. Terrestrial indications are that these exceedingly rare and difficult processes, even in their simplest form, require stable support, such as the proposed clay surfaces on the Earth, and thousands, if not millions, of years of fairly stable temperatures. Even were we to assume that both planets have enough dust in their atmosphere to provide the required stability for the assembly of DNA molecules, the second condition apparently cannot be met. The rapid convection and circulation patterns on these planets—between layers so hot that the components necessary for life would melt or decompose, and others so cold that the materials involved would be solid, and so essentially unreactive—suggests that the necessary temperatures might prevail for continuous periods of only about one day. This would preclude formation of even the simplest living organism on these two planets. A final argument against the existence of life in atmospheres of Venus or Jupiter and Saturn under apparently favorable circumstances is the fact that although our atmosphere is full of microorganisms, all are lifted by winds from levels close to the surface of the Earth and none is of truly atmospheric origin.

What we know about Uranus and Neptune is not very certain but if there are any solid surfaces on these planets, they are in the very deep, hot interior; the rest of both planets is a fluid atmosphere to which all the uncertainties and arguments given above for Jupiter and Saturn are applicable. Pluto's very low temperature and chemically limited environment rule out any possible life on that planet.

Could life exist on satellites of the various planets? Our Moon's surface temperature varies from very cold to very hot, and the Moon has essentially no atmosphere now and probably never had one. The hundreds of kilograms of samples brought from the Moon by the Apollo missions do not indicate any trace of past life on this body. A glimmer of hope for life elsewhere has been raised by the suggestion that below the icy crusts of Jupiter's satellite Europa and perhaps of Saturn's Enceladus, there may exist liquid water. The arguments are, however, still very unconvincing. All other satellites, with the notable exception of Saturn's Titan, also lack a substantial atmosphere, an essential condition for life. Similar arguments apply to the asteroids, because they are too small to hold an atmosphere. There is, of course, the remote possibility that some may have icy interiors with enough heat from radioactive elements to keep part of them liquid and thus perhaps provide an environment suitable for microorganisms.

Titan, Saturn's largest satellite, has long been in the forefront of speculation concerning possible seats of life. As a result of the Voyager investigations, however, our ideas about this satellite have undergone rather drastic changes. The temperature of the surface is much lower than previously expected and far too low for either water or liquid ammonia to be present. On the other hand, the high methane content of its atmosphere may lead to liquid methane pools and to the formation of a whole series of organic compounds. In fact, the presence on Titan of some such compounds containing nitrogen was recently ascertained. The mere presence of organic compounds, however, is a far cry from proving that life exists in Titan's environment. The lower temperatures provide a strong negative indication.

To sum up, we arrive at an important, though in many ways disappointing, conclusion that it is most unlikely that any form of even the most primitive life exists now in the solar system outside the Earth. Whether life of one kind or another did once exist on Mars or on Venus is another question, one that must await further planetary exploration. There is also a non-negligible chance that after the present ice age on Mars is over, life may appear on that planet, although we have no idea how this life would start. If mankind has not yet been obliterated through warfare or uncontrolled pollution by that time there is the possibility that interplanetary travel will have developed to the point where a planned transfer of life from the Earth to Mars may occur.

PLUTO

THE FUTURE

Within the last few decades enormous steps have been taken towards obtaining qualitative and often even quantitative understanding of the universe and our place in it. We can expect a comparable, if not greater, progress in the near future in deciphering the past and the present of our solar system. Further progress is greatly hampered by the inability to make comparisons of our planetary system with other similar systems. We are certain that many other planetary systems exist but at present the observational limitations are formidable. The evolutionary questions—such as how the various galaxies and stars have formed—can be reasonably well answered. We also have a fairly good idea of how our solar system was formed, but many puzzles, such as the striking differences among the planets—and especially the origin of the variety of the satellites—have been only tentatively solved.

The biggest and the most fundamental question is undoubtedly the origin of life and, in particular, of intelligent life in the solar system. As we have seen, the results of the search for it have not been encouraging. There *are* scientists who think that a search for intelligent life, even though it is less likely to exist than primitive life, has a not negligible chance of success. Perhaps in a hundred years, we will know enough to be more optimistic about the outcome of further search than we are today. No doubt the evidence for any kind of life elsewhere would have a profound influence on our ideas concerning our place in this universe and the importance we attach to mankind and its problems: Civilizations may exist that are based on spacial or temporal or philosophical concepts entirely different from ours.

Rational inquiry into the complexity and beauty of the material universe will continue, with its most important goal the strengthening of our intuitive and emotional awareness of mankind's place in this world.

APPENDICES

CALCULATING DISTANCE, MASS, AND SIZE OF SOLAR SYSTEM BODIES

The distance of nearby bodies such as the Moon, Mars, and even asteroids can be calculated by determining simultaneously their position (with respect to fixed stars) at two widely separated points on the Earth. From each point the apparent angle between the body and the star will be different, the nearer the body to the Earth, the greater the difference. Using that difference, one can calculate the distance between the body and the viewer. Radar measurement, which gives more precise data, is based on the time it takes a radio wave to be reflected back to its transmitter. Since radio waves travel at a known speed (the speed of light), the distance to the reflecting body can be calculated directly. For more distant bodies Kepler's third law—which states that the period of revolution of a planet has a known relation to its distance from the Sun—may be applied: thus, knowing the period—more or less readily observed—we can calculate the distance.

The diameter of a planet, once we know its distance, can be measured from its image on a photograph; the distance of a satellite from a planet and sometimes its size can also be determined in the same manner. Occultations of stars by planets or satellites—that is, the interruption of the light from the star as the body passes between us and the star—also offers a means of determining size where we know the speed of the body.

The mass of a planet can be calculated from the period of revolution of a satellite and Kepler's third law. Where the planet has no satellites, mass can be determined by observation of the manner in which the planet perturbs the motion of other planets.

MEASUREMENT, NOTATION, AND BASIC CONSTANTS

METRIC SYSTEM

		Length	Fraction or Multiple of the Basic Unit	Mass/Weight	
		micrometer	1/1,000,000	microgram	
		millimeter	1/1,000	milligram	
		centimeter	1/100	centigram	
English equivalent	39.4 inches	METER (basic unit)	1	GRAM (basic unit)	.035 ounces *English equivalent*
	.621 miles	kilometer	1,000	kilogram	2.21 pounds

TEMPERATURE SCALES

In the Fahrenheit (F) system, the freezing point of water is 32° and the boiling point, 212°. In the Centigrade or Celsius (C) system, the freezing point is 0° and the boiling point 100°, so that the Celsius degree subdivisions are 1.8 times larger than those of Fahrenheit. The Kelvin (K) system, whose subdivisions are the same as those of Celsius, has its lower limit at absolute zero (-273° C and -459° F), where all molecular motion stops; hence the freezing point of water in the Kelvin system is 273° and the boiling point, 373°.

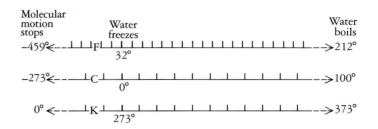

BASIC CONSTANTS

Astronomical unit (AU)	149.6 million kilometers (The mean distance from the Earth to the Sun.)
Light year (ly)	9.46×10^{12} kilometers, or about 10,000 billion kilometers (The distance that light travels in one year.)
The speed of light (c)	299,792.5 kilometers per second
Mass of the Sun ($M_\odot$)	1.999×10^{33} grams, or about 2,000 billion billion billion kilograms
Mass of the Earth ($M_\oplus$)	5.97×10^{27} grams, or about 6 million billion billion kilograms

SCIENTIFIC NOTATION:
QUANTITIES EXPRESSED AS
POWERS OF TEN

Writing out very long numbers can be cumbersome and time-consuming. Scientific notation avoids this by expressing such numbers as a number (less than 10) multiplied by some exponential value of ten. Thus, 8,000,000 would be expressed as 8×10^6 ($10^6 = 10 \times 10 \times 10 \times 10 \times 10 \times 10$); 120,000,000 would be 1.2×10^7; and 9,460,000,000,000 (the length of a light year in kilometers) would be 9.46×10^{12}.

PROPERTIES OF THE PLANETS

	Mercury	Venus	Earth	Mars	Jupiter	Saturn	Uranus	Neptune	Pluto
Distance from Sun (millions of kilometers)									
Maximum	69.7	109	152.1	249.1	815.7	1,507	3,004	4,537	7,375
Minimum	45.9	107.4	147.1	206.7	740.9	1,347	2,735	4,456	4,425
Mean	57.9	108.2	149.6	227.9	778.3	1,427	2,869.6	4,496.6	5,900
Period of revolution	88 days	224.7 days	365.26 days	687 days	11.86 years	29.46 years	84.01 years	164.8 years	247.7 years
Period of rotation	59 days	−243 days‡	23 hr. 56 min. 4 sec.	24 hr. 37 min. 23 sec.	9 hr. 50 min. 30 sec.	10 hr. 14 min.	−11 hr.‡	16 hr.	6 days 9 hr.
Equatorial diameter (kilometers)	4,880	12,104	12,756	6,787	142,800	120,000	51,800	49,500	5,000 (?)
Mass (Earth = 1)	.055	.815	1	.108	317.9	95.2	14.6	17.2	.003 (?)
Density (Water = 1)	5.4	5.2	5.5	3.9	1.3	.7	1.2	1.7	<1
Atmosphere*	None	Carbon Dioxide	Nitrogen, Oxygen	Carbon Dioxide, Argon (?)	Hydrogen, Helium	Hydrogen, Helium	Hydrogen, Helium, Methane	Hydrogen, Helium, Methane	Methane (?)
Mean temperature at visible surface (degrees celsius)†	350(S) day, −170(S) night	−33 (C) 480 (S)	22 (S)	−23 (S)	−150 (C)	−180 (C)	−210 (C)	−220 (C)	−230 (?)
Surface gravity (Earth = 1)	.37	.88	1	.38	2.64	1.15	1.17	1.18	?
Inclination of Axis	<28°	3°	23°27′	23°59′	3°05′	26°44′	82°5′	28°48′	?
Inclination of orbit to ecliptic	7°	3.4°	0°	1.9°	1.3°	2.5°	.8°	1.8°	17.2°

*Main components.
†S indicates solid; C, clouds.
‡Retrograde movement.

PROPERTIES OF SATELLITES

Satellite	Distance from planet (kilometers)	Period of revolution (days)	Radius (kilometers)
of the Earth			
The Moon	385,000	27.32	1738
of Mars			
Phobos	9,380	0.32	14
Deimos	23,500	1.26	8
of Jupiter			
Amalthea	181,000	0.50	80
Io	422,000	1.77	1830
Europa	671,000	3.55	1550
Ganymede	1,070,000	7.16	2640
Callisto	1,880,000	16.69	2500
Leda	11,110,000	239.0	8
Himalia	11,500,000	250.6	60
Elara	11,700,000	259.7	20
Lysithea	11,900,000	263.6	7
Ananke	21,200,000	631.1*	6
Carme	22,600,000	692.5*	7
Pasiphae	23,500,000	738.9*	6
Sinope	23,700,000	758.0*	7
of Saturn			
Janus	169,500	0.74	100
Mimas	186,000	0.94	200
Enceladus	238,000	1.37	300
Tethys	295,000	1.89	500
Dione	377,000	2.74	400
Rhea	527,000	4.52	750
Titan	1,220,000	15.95	2900
Hyperion	1,480,000	21.28	200
Iapetus	3,560,000	79.33	750
Phoebe	13,000,000	550.5*	100
of Uranus			
Miranda	130,000	1.41	200
Ariel	191,000	2.52*	700
Umbriel	260,000	4.14*	500
Titania	436,000	8.71*	900
Oberon	583,000	13.46*	800
of Neptune			
Triton	354,000	5.88*	1900
Nereid	5,570,000	359.4	300

*Retrograde movement.

RECOMMENDED READING

BASH, FRANK N. *Astronomy.* Harper, 1977.
    A particularly good introduction for readers with no specialization in science.

HARTMANN, W.K. *Moons and Planets: An Introduction to Planetary Science,* ed. 2. Wadsworth, 1983.
    An excellent, authoritative text that employs negligible amounts of mathematics.

JASTROW, ROBERT. *Red Giants and White Dwarfs.* Harper, 1967.
    An excellent, particularly readable treatment of its topic.

ROBBINS, R.R., AND HEMENWAY, MARY K. *Modern Astronomy: An Activities Approach.* Univ. Texas Press, 1982.
    An up-to-date introduction for readers with a non-technical background that includes applications and exercises useful for amateur astronomers.

SNOW, T.T. *The Dynamic Universe.*
    A very recently published textbook for non-science majors.

WHIPPLE, FRED L. *Orbiting the Sun: Planets and Satellites of the Solar System.* Harvard Univ. Press.
    A readable and informed description of its topic with excellent illustrations.

INDEX

page 74
NASA–JPL.

page 75 (top left)
NASA–JPL.

page 75 (top left and bottom)
NASA–NSSDC.

page 77 (top and bottom)
NASA.

page 78
NASA–JPL.

page 79 (left and right)
NASA.

page 81
NASA–JPL.

page 84
NASA–JPL.

page 86
NASA–JPL.

page 87
NASA–JPL.

page 88
NASA–JPL.

page 91
NASA–JPL.

page 92
NASA–JPL.

page 93
NASA–NSSDC.

page 96
NASA–JPL.

page 97 (left)
NASA–JPL.

page 97 (right)
NASA–NSSDC.

page 98 (left)
NASA–NSSDC.

page 98 (right)
NASA–JPL.

page 99 (left)
NASA–JPL.

page 99 (right)
NASA–NSSDC.

page 101
NASA–JPL.

page 102 (top and bottom)
NASA.

page 103
NASA.

page 104
NASA.

page 106
NASA–JPL.

page 107 (left and right)
NASA–JPL.

page 108 (top)
NASA–JPL.

page 108 (bottom)
NASA.

page 109
NASA–JPL.

page 110
NASA.

page 111
NASA.

page 113
Ron Miller.

page 115 (top and bottom)
Lick Observatory.

page 116
Project Stratoscope II, Princeton University,
supported by NSF and NASA.

page 119
Redrawn from the Astronomical Journal,
"Rings of Uranus: Results of April 10,
1978 Occultations," Nichelson, P. D. et al,
83, p. 1248 (1978).

page 122
Palomar Observatory, California Institute of
Technology.

page 123
U.S. Naval Observatory.

page 127
E. F. Helin, Palomar Observatory.

page 129 (top and bottom)
Mount Wilson and Las Campanas Observatories, Carnegie Institution of Washington.

page 130
The Bettmann Archive.

page 133 (top left and right)
Donald E. Brownlee.

page 133 (bottom)
Palomar Observatory, California Institute of
Technology.

page 135 (left)
Sovfoto.

page 135 (right)
National Geographic Society.

page 136 (bottom)
The Naval Research Laboratory.

page 137 (top right)
James Baker.

page 137 (bottom left)
Donald E. Brownlee.

page 138 (top)
NASA–Lyndon B. Johnson Space Center.

page 138 (bottom)
U.S. Geological Survey.

page 139
Dave McLean, Lunar and Planetary Laboratory.

page 141
Stephen Murray and colleagues at the Center
for Astrophysics of the Harvard College
Observatory and the Smithsonian
Astrophysical Observatory.

page 142
Copyright ©, California Institute of Technology.

page 147 (top)
Nelson Max, Lawrence Livermore Laboratory.

page 147 (bottom)
H. S. Forrest.

page 149
Sigurgeir Jonasson.

page 151 (left)
G. Penso.

page 151 (right)
Carl T. Mattern, N. I. H.

page 152 (top right)
S. M. Awramik.

page 152 (top right, middle and bottom
right)
J. William Schopf.